S/O ML

CIRIA C690

London, 2010

WaND

Guidance on water cycle management for new developments

D Butler	University of Exeter
FA Memon	University of Exeter
C Makropoulos	National Technical University of Athens
A Southall	University of Exeter
L Clarke	CIRIA

Books are to be returned on or before
the last date below.

Classic House, 174–180 Old Street, London EC1V 9BP
TEL: 020 7549 3300 **FAX:** 020 7253 0523
EMAIL: enquiries@ciria.org **WEBSITE:** www.ciria.org

WaND Guidance on water cycle management for new developments

Butler, D, Memon, FA, Makropoulos, C, Southall, A, Clarke, L

CIRIA C690 © CIRIA 2010 RP777 ISBN: 978-086017-690-9

British Library Cataloguing in Publication Data

A catalogue record is available for this book from the British Library.

Keywords		
Sustainable water cycle management, greywater, rainwater harvesting, SUDS, water efficiency, stakeholder engagement, sustainability, water and new developments (WAND) water management		
CIRIA Themes		
Sustainable water management, flood risk management and surface water drainage, sustainability and the built environment		
Reader interest	**Classification**	
Land-use planning, water industry, water supply, water resources, water use, environmental regulation, provision and maintenance of sustainable water management systems	Availability	Unrestricted
	Content	Advice/guidance, original research
	Status	Author's opinion, committee-guided
	USER	Land-use planners, water industry, environmental regulators, developers

Published by CIRIA, Classic house, 174-180 Old Street, London, EC1V 9BP

This publication is designed to provide accurate and authoritative information on the subject matter covered. It is sold and/or distributed with the understanding that neither the authors nor the publisher is thereby engaged in rendering a specific legal or any other professional service. While every effort has been made to ensure the accuracy and completeness of the publication, no warranty or fitness is provided or implied, and the authors and publisher shall have neither liability nor responsibility to any person or entity with respect to any loss or damage arising from its use.

All rights reserved. No part of this publication may be reproduced or transmitted in any form or by any means, including photocopying and recording, without the written permission of the copyright-holder, application for which should be addressed to the publisher. Such written permission must also be obtained before any part of this publication is stored in a retrieval system of any nature.

If you would like to reproduce any of the figures, text or technical information from this or any other CIRIA publication for use in other documents or publications, please contact the Publishing Department for more details on copyright terms and charges at: publishing@ciria.org or tel: 020 7549 3300.

For further information about CIRIA publications go to: <www.ciria.org>

WITHDRAWN FROM LIBRARY STOCK
UNIVERSITY OF STRATHCLYDE
12 APR 2010
UNIVERSITY LIBRARY
D
628.1094'1
WAN

Executive summary

This publication provides guidance on how to achieve improved sustainable water cycle management in new developments. It outlines the policy context and the processes involved in sustainable water cycle management focusing on why it is required and what approaches are needed for its successful planning and delivery.

The research for the guidance was conducted by the WaND (water cycle management for new developments) research project whose aim was to support the delivery of a more integrated and sustainable approach to water management for new developments by the provision of tools and guidelines for project design, delivery and management. This guidance is based on the information collated in this multi-disciplinary research and presents the results for practitioners involved in water cycle management in new developments.

Increases in the demand for water coupled with climate change and demographic shifts means that water stress is a growing problem in the UK. This has highlighted that traditional methods of water cycle management are unsustainable and it is increasingly important to consider alternative approaches. This document provides practitioners with an improved understanding of the tools and techniques necessary for achieving, delivering and adopting sustainable water cycle management in new developments.

The changing perception of the water environment and the dynamic policy, social and technical responses have led to an increasingly more sustainable approach to water management. Although sustainability is not a set outcome the concept has been discussed in relation to the water cycle. This includes new technologies for delivering sustainable water cycle management to conserve, reuse and manage water in the environment.

The technologies explored include water-saving devices, rainwater harvesting, greywater recycling and sustainable drainage systems (SUDS). Operational or research tools were developed in the WaND research project to evaluate performance and provide evidence on the suitability of these technologies. The tools identify important design and performance criteria/indicators and determine their potential contribution to sustainability. The tools that were developed primarily for the WaND research project are based on current and innovative water management technologies and practices. The guidance also highlights the importance of stakeholder engagement from an early stage to ensure support in the delivery of sustainable water cycle management.

There are clear drivers for improved water management but a lack of knowledge on approaches, available technologies, delivery and the costs and maintenance of the systems has resulted in their slow uptake. This guidance addresses these issues and provides recommendations to encourage the use of sustainable water cycle management.

The planning process is central in achieving sustainable water cycle management and it is recommended that water supply, wastewater disposal and surface water management are considered early in the planning process. Water-saving appliances can contribute to considerable water savings and several devices are available including low-flush toilets and aerated showers. Rainwater harvesting and greywater systems have the potential to save large amounts of water but need to be applied on the appropriate application scale because of their size and associated costs. Rainwater harvesting systems are encouraged in large buildings and in wetter parts of the country. The supply from greywater systems is reliable because of the water availability and is recommended for multi-dwellings, blocks of flats, hotels and large public and commercial buildings. However they require more extensive treatment processes that vary in complexity and acceptability.

The WaND research project tested five different greywater technologies to establish their performance, cost of operation and water quality standards and it was accepted that there were other systems on the market. It was also acknowledged that SUDS are an important part of surface water management and should be considered instead of traditional drainage methods providing local benefits in terms of flood risk management, water availability and quality of life.

Other recommendations include a change in peoples' perception of water and the relationship between water, energy use and consequently carbon reduction. Water should have greater value, which could result in less waste.

Acknowledgements

Authors

David Butler BSc MSc PhD DIC CEng CEnv FICE FCIWEM

David Butler is professor of water engineering at the University of Exeter and director of the Centre for Water Systems. He specialises in urban water management and was the director of the WaND research project. Prof Butler has authored/co-authored over 200 technical papers and several books/published reports including *Urban drainage* and *Water demand management*. He is a chartered civil engineer and fellow of the Institution of Civil Engineers (ICE) and Chartered Institution of Water and Environmental Management (CIWEM). He is editor of CIWEM's *Water & and Environment Journal* and the *Urban Water Journal*.

Fayyaz Ali Memon BE (Civil) MA MSc DIC PhD

Dr Fayyaz Ali Memon is a lecturer in water engineering at the University of Exeter. He specialises in sustainable urban water management. He was responsible for managing the WaND reasearch project and contributing to research activities on water recycling and wastewater collection systems. He has over 60 publications to his credit, including three books on various aspects of urban water management. He is a chartered engineer and environmentalist, member of the Chartered Institution of Water and Environmental Management (CIWEM) and the co-ordinator of the WATERSAVE Network.

Christos Makropoulos MEng (Civil) MSc DIC PhD

Dr Christos Makropoulos is a lecturer in the School of Civil Engineering at the National Technical University of Athens. His research focuses on decision support systems in urban water management and water resources applications. He was responsible for the decision support tools development within the WaND research project. Dr Makropoulos has authored more than 70 journal and conference papers, best practice guidelines and book chapters, and is an associate editor of the *Urban Water Journal* and a visiting fellow of the University of Exeter.

Andrew Southall BSc MSc DIC CGeol FGS CWEM FCIWEM FRSA

Andrew qualified from Imperial College in 1981. He started as an assistant marine scientist working on surveys for long sea outfalls and after some time in marine geophysics (which included the marine geophysical survey for the channel tunnel) he worked first for WRc and then for Ewan Group. Andrew holds an honorary research fellowship at the Centre for Water Systems at the University of Exeter.

Louise Clarke BA MArch

Louise Clarke is an assistant project manager at CIRIA involved with projects primarily related to water management. Before this she completed a masters in urban design at the University College London looking at the delivery of green infrastructure and sustainable drainage through the master planning and spatial planning process.

Steering group

Following CIRIA's usual practice, the research project was guided by a steering group:

David Balmforth (chair)	MWH
Justin Abbott	Arup
Alistair Atkinson	WSP
David Balmforth	MWH
Steve Ball/Owen Peat	Homes and Communities Agency
Peter Bide	DCLG
Phil Chatfield	Environment Agency
Jonathan Dennis	Environment Agency
Sian Hills	Thames Water
Paul Jeffrey	Cranfield University
Peter Jiggins	Defra
Sian Lewis	Home Builders Federation
Andy McConkey	Halcrow
Alex Nickson	Greater London Authority
Liz Sharp	Bradford University

CIRIA managers

CIRIA's project managers for the guide were Robin Farrington and Paul Shaffer.

Project funders

The project to produce the guidance was funded by:

EPSRC
Homes and Communities Agency
Environment Agency

The WaND research project was funded by:

EPSRC
Environment Agency
Severn Trent Water
Yorkshire Water
Thames Water
Three Valleys Water
Essex and Suffolk Water

Contributors

CIRIA wishes to acknowledge the following for providing substantial further information:

Jonathan Dennis	Environment Agency
Owen Peat	Homes and Communities Agency
Sian Hills	Thames Water
Andy McConkey	Halcrow

Contents

Executive summary iii

Acknowledgements v

Boxes ix

Case studies x

Figures x

Tables xi

Glossary xiii

Abbreviations xx

1 Introduction 1

1.1 The water cycle 1
1.2 Aim of this guidance 2
1.3 The WaND research project 2
1.4 Context of the document 2
1.5 Who should read this guidance? 3
1.6 Structure of the guidance 3

2 Water cycle management in context 5

2.1 Introduction 5
2.2 Background 5
2.3 Pressures resulting in policy change 6
2.3.1 Environmental pressures 6
2.3.2 Social and political pressures 8
2.4 Policy pressures 9
2.4.1 EU Directives 9
2.5 Responses 11
2.5.1 Policy and regulatory responses 11
2.5.2 Research responses 18
2.5.3 Implementation responses 19
2.6 Carbon reduction and the water cycle 21
2.7 Conclusions 22

3 Planning for water 23

3.1 Introduction 23
3.2 Challenges 23
3.3 What is sustainability? 24
3.4 How is sustainability linked to water cycle management? 25
3.5 How can sustainable water cycle management be achieved? 26
3.5.1 Unsustainable development 26
3.5.2 Sustainable development 26
3.5.3 Sustainability and water planning 28
3.5.4 Sustainability: principles, criteria, indicators 30

3.6 Water management decisions for the future 32

3.6.1 Using the scenarios 33

3.7 Conclusions 36

3.9 Further research and guidance 36

4 Technologies 37

4.1 Introduction 37

4.2 Water demand management 38

4.2.1 Water saving devices 38

4.2.2 Rainwater harvesting (RWH) 40

4.2.3 Greywater reuse 42

4.3 Stormwater management 45

4.3.1 SUDS 46

4.3.2 RWH as a stormwater management option 52

4.4 Conclusions 52

4.5 Further research and guidance 54

5 Stakeholder engagement 55

5.1 Introduction 55

5.2 Learning through case studies 55

5.3 Descriptions of the case studies 55

5.4 Practical delivery of SWCM 58

5.4.1 Planning policy and strategic implementation 58

5.4.2 Incentives 60

5.4.3 Reliability of new systems 60

5.4.4 Acceptance of SWCM measures 60

5.5 Effective public engagement 61

5.5.1 The need for engagement 61

5.5.2 Understanding end user behaviours and perceptions 61

5.5.3 Framing effective messages 64

5.5.4 Guidance for good quality engagement involvement at a local scale 64

5.6 Taking up SWCM in the future 65

5.6.1 Organisational methods required 65

5.6.2 Recommendations for taking up SWCM 66

5.7 Conclusions 67

5.8 Further research and guidance 68

6 Tools for decision making 69

6.1 Introduction 69

6.2 The WaND decision support tools 69

6.2.1 Project assessment tool (PAT) 70

6.2.2 Site screening tool (SST) 71

6.2.3 Urban water optioneering tool (UWOT) 73

6.2.4 Suitability evaluation tool (SET) 76

6.2.5 Demand forecasting tools 77

6.2.6 Stormwater management tools 78

6.2.7 Greywater recycling tools 80

6.2.8 Health impact assessment tool 82

6.3 Use and development of the tools 83
6.4 Conclusions 86
6.5 Further research and guidance 86

7 Final conclusions 87
7.1 Recommendations 87
7.2 Key messages 89

8 References 91
Acts and Bills 101
British Standards 101
Regulations 101

Appendices 103

A1 The WaND research project and portal 103

A2 Further discussion on sustainability 105

A3 Water cycle studies 107
A3.1 Water cycle studies 107
A3.2 Scenarios for water cycle management 111

A4 WaND research on water management technologies 113
A4.1 Technologies available for SWCM 113
A4.2 Guidance reports on suitable water management technologies 117
A4.3 Water saving devices 118
A4.4 WaND research – low-flush toilets 121
A4.5 Rainwater harvesting 124
A4.6 Greywater 126
A4.7 Stormwater management 132

A5 List of collaborators involved in the WaND research 135

A6 LANDCOMs list of treatment technologies and their associated attributes 138

A7 Performance of the tested greywater technologies in the WaND research 142

A8 Contact details of organisations involved in developing the decision support details 143

Boxes

Box 4.1 RWH as water demand management (WDM) option – scale of use 40
Box 4.2 Stormwater drainage performance sustainability evaluation methodology testing at Elvetham Heath 49
Box 4.3 SUDS field monitoring – preliminary observations 50
Box 4.4 SUDS implementation in the Dunfermline Eastern Expansion Area (DEX) 51
Box 5.1 Sheffield case study – SUDS 56
Box 5.2 Childwall, Liverpool case study – water saving devices, rainwater harvesting and greywater recycling 57
Box 5.3 Elvetham Heath case study – water saving devices and SUDS 57
Box 5.4 Supplementary study on user interactions with SWCM strategies 63
Box 6.1 Using the site screening tool in the Humber sub-region 73
Box 6.2 UWOT application in Elvetham Heath 75

Box A2.1 The journey ahead: into the unknown 106
Box A4.1 Some guidance reports on sustainable water management technologies 117
Box A4.2 Small-bore wastewater collection system field trials – WRc case study .123
Box A4.3 Examples of RWH systems installed in the UK 124
Box A4.4 Pilot scale MBR system – construction and operating conditions 126
Box A4.5 Pilot scale MCR system – construction and operating conditions 127
Box A4.6 Pilot scale VFRB and HFRB system – construction and operating conditions 127
Box A4.7 Pilot scale GROW system-construction and operating conditions 128
Box A4.8 Examples of greywater recycling systems used for economic assessment 131
Box A4.9 SUDS documents 132

Case studies

Case study 3.1 Millennium Green, Nottinghamshire 27
Case study 3.2 Elvetham Heath, Hampshire 27
Case study 3.3 Dunfermline Eastern Expansion, Scotland 28

Figures

Figure 2.1 Areas of relative water stress 8
Figure 2.2 Possible eco town locations 20
Figure 2.3 Thames Gateway 21
Figure 3.1 Shared UK principles of sustainable development 25
Figure 3.2 Scenario comparisons in terms of water (as per cent improvement from benchmark) 35
Figure 4.1 Detention ponds at Elvetham Heath 49
Figure 4.2 Swale at Evetham Heath 49
Figure 4.3 Pond at Elvetham Heath 49
Figure 4.4 Wetland area 51
Figure 4.5 Detention basin 51
Figure 4.6 Detention basin before a storm event 51
Figure 4.7 Detention basin after a storm event 51
Figure 5.1 SUDS from Sheffield case study 56
Figure 5.2 Wetland from Sheffield case study 56
Figure 5.3 Communal rainwater recycling system pump room at Childwall 57
Figure 5.4 Individual household rain and greywater recycling underground tanks from Childwall 57
Figure 5.5 SUDS at Elvetham Heath 57
Figure 5.6 Pond at Elvetham Heath 57
Figure 6.1 Map of Humber estuary showing suitability areas 74
Figure 6.2 A comparison of the water save scenario with the benchmark scenario against multiple sustainability objectives 76
Figure 6.3 Example output from LCA tool showing environment effect of materials used in membrane based technologies 81
Figure 6.4 Mean DALY scores for each of the selected water management options (on a log scale) in comparison with the WHO and screening reference levels 83

Figure 6.5 WaND tools interaction .85
Figure A2.1 The Spirit of St Louis .106
Figure A3.1 Different stages of a water cycle study .109
Figure A3.2 How this guidance relates to the WCS guidance .110
Figure A3.3 The relationship between sustainability principles, criteria and indicators .111
Figure A3.4 The Foresight Future .112
Figure A4.1 Effect of flush volume (100 mm pipe, gradient 1:100) .122
Figure A4.2 Comparison of the performance of the ULFT and a dual-flush WC . .123
Figure A4.3 Chemical system (MCR) schematic .127
Figure A4.4 Pilot scale reed bed .127
Figure A4.5 GROW system construction .128
Figure A4.6 GROW system operation phase .129
Figure A4.7 Environmental effect of different types of SUDS .133
Figure A4.8 Environmental effects of SUDS and conventional drainage .133
Figure A4.9 Runoff reduction in volume from roads and buildings with a storage tank of 0.75m³/person .134
Figure A4.10 Runoff reduction in volume from roads and buildings with a storage tank of 1.5m³/person .134

Tables

Table 3.1 Achieving the Code for Sustainable Homes .29
Table 3.2 Categories and primary criteria of sustainability evaluation used by SWARD .31
Table 3.3 Categories and primary criteria for a sustainable appraisal used by SPeAR® .32
Table 3.4 Objectives for SWCM .34
Table 3.5 Distributed infrastructure scenarios .35
Table 4.1 National planning policies for SUDS promotion .46
Table 4.2 SUDS performance .47
Table 5.1 Checklist for framing effective messages .64
Table 5.2 Recommendations for SWCM uptake .66
Table 6.1 WaND decision support tools .70
Table 6.2 Categories and primary criteria of sustainability evaluation used by PAT . .71
Table 6.3 Environmental sustainability criteria and indicators .72
Table 6.4 Social sustainability criteria and indicators .72
Table 6.5 Economic sustainability criteria and indicators .72
Table 6.7 Demand forecasting tools developed in WaND .78
Table 7.1 Key messages for stakeholders .90
Table A3.1 Stages of the water cycle studies .108
Table A4.1a Technologies available for water supply management .113
Table A4.1b Technologies available for stormwater management .114
Table A4.1c Technologies available for wastewater management .115
Table A4.1d Technologies available for water recycling/reuse .116
Table A4.2 Water consumption using conventional appliances .118
Table A4.3 Estimate of potential water saving with best available technologies not entailing excessive costs (BATNEEC) and estimated future technologies .118

Table A4.4 Expected demand reduction effects of water efficiency options based on their performance and likely uptake in new build and existing housing stock . . . 119
Table A4.5 Potential water savings and CO_2 emissions through shower use . . . 120
Table A4.6 Suitability, advantages and disadvantages of water-saving taps . . . 120
Table A4.7 Suitability, advantages and disadvantages of WCs flush volume reduction approaches . . . 121
Table A4.8 Water and energy savings with ULFT s (relative to conventional WC) . . 124
Table A4.9 RWH system maintenance activities and corresponding frequency . . . 125
Table A4.10 Cost-benefit analysis of RWH systems described in Box 4.4 . . . 125
Table A4.11 Examples of two rainwater harvesting schemes with short payback periods . . . 125
Table A4.12 Critical concentrations of pollutants affecting microbial activity . . . 126
Table A4.13 Plants in the GROW system . . . 128
Table A4.14 Compliance level of the systems for different standards and guidelines . . . 129
Table A4.15 The characterised effect (development scale: 500 households) . . . 130
Table A4.16 Energy consumption of the investigated technologies . . . 130
Table A4.17 Total capital cost for the small and large-scale example systems . . . 131
Table A 4.18 Operation and maintenance cost and savings . . . 132
Table A5.1 Organisation and collaborators involved in the WaND research . . . 135
Table A5.2 Others involved in WaND . . . 136
Table A7.1 Performance of the five greywater treatment technologues tested in the WaND research project . . . 142

Glossary

Abstraction	The removal of water from any source, either permanently or temporarily.
Aquifer	A geological formation, group of formations or part of a formation that can store and transmit water in significant quantities.
Attenuation	Reduction in flow through natural or artificial storage that increases the duration of flow hydrology.
Borehole	Narrow well hole sunk into a water-bearing rock from which water may be pumped or the groundwater level measured.
Biodiversity	The variety of life in all its forms, levels and combinations. Includes ecosystem diversity, species diversity and genetic diversity.
Brownfield site	Land that has been developed in the past (see previously developed land).
Catchment	The area from which precipitation and groundwater will collect and contribute to the flow of a specific river.
Climate change	Climate change refers to the long-term variations in global temperature and weather patterns. Climate change specifically refers to the effect on the climate by both natural and human activity, primarily greenhouse gas emissions.
Combined sewer	A sewer designed to carry foul sewage and surface water runoff in the same pipe.
Commuted sum	A sum of money paid in the present, from a developer to the organisation adopting a development to cover the costs associated with the maintenance of aspects of the development.
Continuous simulation	A single long-duration hydraulic simulation using a continuous record of rainfall (eg one year or five years).
Contributing area	The area that contributes storm runoff directly to the sewage system.
Conventional toilet	Toilet pan coupled to a flushing cistern and flushing device, used to dispose of urine and faecal waste.
Decision map	Conceptual illustrations of decisions made in practice based on case studies. Used as a means to provide bases for discussion and inclusion of sustainability considerations in water management decisions.
Decision support system	An interactive flexible, and adaptable computer-based information system, especially developed for supporting the solution of a non-structured management problem for improved decision making. It uses data, provides an easy-to-use interface, and allows for the decision maker's own insights.

Demand forecasting tool	A strategic water management tool providing forecast of future water needs. It provides a set of scenarios for stakeholders by taking into account influencing factors on water demand including population growth, lifestyle, climate change and household size.
Demand management	The delivery of policies or measures that serve to control or influence the consumption or waste of water (this definition can be applied at any point along the chain of supply).
Drought order	A means whereby water companies or the Environment Agency can apply to the secretary of state or the Welsh Assembly Government for the imposition of restrictions in the uses of water (and/or) that allows for the abstraction of water outside the existing licence conditions.
Effluent	Liquid waste from industrial, agricultural or sewage plants.
Evapotranspiration	The sum of evaporation and plant transpiration. In the context of this WaND research it is used in the calculation of NAPI. Potential Evapotranspiration (PET) is a representation of the environment demand for evapotranspiration and is used in the volumetric infiltration model developed for the hydraulic performance assessment tool.
Event (rainfall)	Single occurrence of a rainfall period before and after which there is a sufficient dry period for runoff and discharge from the drainage system to cease.
Floodplain	Low-lying land adjacent to the river over which the river flows at time of flood, and which has been formed by channel migration and deposition of alluvial sediments.
Flush volume	Volume of water released with one flush of WC.
Flush wave	Wave resulting from WC flushing, consisting of a large volume of water in a short period of time.
Government office region	Highest level of English local government geography comprising of nine regions.
Greenfield	Land that has never been developed, other than for agricultural or recreational use.
Greenfield runoff rate	The rate of runoff that would occur from the site in its undeveloped (and undisturbed) state.
Greywater	Wastewater from sinks, baths, showers and domestic appliances (excluding kitchen sinks, dishwashers).
Greywater recycling tool	A research-based tool for the WaND researchers. Several greywater tools were developed to investigate aspects related to the environmental, economic and design performance of greywater systems to increase confidence in system performance and specify the appropriate scale of provision.
Greywater system	Water is collected from showers, baths and wash basin. This water is treated for reuse for purposes that do not require drinking water quality.
Gross solids	Sewage-derived material greater than 6 mm in any two dimensions and with specific gravity close to unity. Includes faecal stools, toilet paper, sanitary towels, tampons, condoms,

	and other material that enters public sewers, such as bathroom litter and the products of food waste-disposal units.
Groundwater	Water within the subsurface saturated zone (including aquifers).
Health	"A state of complete physical mental and social well-being and not merely the absence of disease or infirmity" (WHO, 1964).
Health hazard	Anything that can potentially cause harm (with harm being loss of life, injury, illness).
Health impact assessment	A combination of procedures, methods and tools by which a policy, programme or a project may be judged as to its potential effects on the health of a population, and the distribution of those effects within the population.
Health impact assessment tool	A generic methodology that was developed in the WaND research for health impact assessments on innovative water management strategies and technologies.
Health risk	A health risk is a measure of the probability that a health hazard will actually cause harm to a particular individual or group of people.
Hippo	Toilet cistern-displacement device.
Impermeable area	A measure of the surface which resist the infiltration of water usually a roof or road.
Impermeable surface	Surface that resists the infiltration of water.
Infiltration	The passage of surface water through the surface of the ground.
Land-use	The main activity that takes place on an area of land based on economic, geographic or demographic use, such as residential, industrial, agricultural or commercial.
Level of service	The performance of the drainage system with respect to sewer flooding.
Limiting solids transport distance	Maximum distance where a solid will become stationary. Depends on the diameter of the pipe and shape of the solids. The distance is determined when the solid halts its movement after three consecutive flushes.
MACROWater	A scenario-based, top-down model that was part of the demand forecasting tool. It is based on government targets for house building and water efficiency and Ofwat targets for metering. These statistics combined with household and population projections provide a set of alternative futures to consider using urban water management scenarios. They are based on different assumptions about the adoption of water-saving technologies and different water consumption patterns.
Outfall	The point, location or structure where wastewater or drainage discharges from a pipe, channel, sewer drain or other conduit.
Output area	Census output geography introduced to the UK in the 2001 census. Each OA comprises of about 200 persons.
Overflow	The flow of excess water from a storage area when the capacity of that storage is exceeded.

Peak flow	The point at which the flow of water from a given storm event is at its highest.
Percentage runoff	The percentage of the rainfall volume falling on a specified area that enters the stormwater drainage system or discharges to a watercourse (if considering rural percentage runoff).
Permeable surface	A surface formed of material that is impervious to water but, by virtue of voids formed through the surface, allows infiltration through the pattern of voids, for example concrete block paving.
Pervious area	Area of ground that allows infiltration of water, although some surface runoff may still occur.
Pervious pavement	Pavement designed to reduce imperviousness, which minimises surface runoff. They vary in type from porous asphalt, porous concrete or modular paving, and are suited to lightly trafficked areas. Runoff infiltrates to an underlying stone-filled reservoir, which is capable of removing pollutant, before discharge in a controlled manner into a nearby watercourse or directly to the ground. For the purpose of the WaND research discharge is to the ground. Also known as *permeable pavement*.
Previously developed land	Land that is, or was, occupied by a permanent structure (excluding agricultural or forestry buildings) and associated fixed surface infrastructure, including the curtilage of the development.
Project assessment tool	A tool developed for multiple stakeholders to discuss the issues of sustainability through a set of criteria and indicators. The tool is intended for the preliminary stages of project development and to be continuously consulted throughout the delivery of the project.
Rainfall intensity	Amount of rainfall occurring in a unit of time generally expressed in mm/hr.
Rainwater harvesting	The collection of water from roofs or clean hard standing for use for domestic purposes, such as toilet flushing, garden watering and car washing without needing to be cleaned or treated.
Recharge	Replenishment of groundwater storage in an aquifer.
Recycling (water)	The use of water for a second purpose, once it has undergone treatment or recovery processes.
Resources zone	The supply area for water companies where the water resources such as reservoirs, boreholes and rivers are located.
Retention pond	Permanent water body that holds water for sufficient time to allow particles to settle and to provide biological treatment. Retention ponds are known as regional controls, serving large scale developments such as industrial estates and major housing developments.
Return period	A term used to express the frequency of extreme events. It refers to the estimated average time interval between events of a given magnitude (the reciprocal of the annual exceedance probability). In this guidance return period is used to describe extreme rainfall.

Reuse	The use of water that has already been used without treatment.
Runoff	Water that flows over the ground surface to a drainage system. This occurs if the ground is impermeable or if permeable ground is saturated.
Scenarios	Scenarios are descriptions of alternative hypothetical futures that reflect different perspectives on past, present and future developments, which can serve as a basis for action.
Sewer flooding	The blockage or overflowing of the stormwater drainage system causing it to flood.
Site screening tool	The tool establishes the "best" location for new urban developments based on the primary criteria of sustainability (economic, environmental, social and technical). It uses a multi-criteria approach to prioritise suitable sites and screen out unsuitable sites using GIS software.
Soakaway	A substrate structure into which surface water is conveyed, designed to promote infiltration.
Soil moisture deficit	A measure of soil wetness, calculated by the Meteorological Office in the UK, to indicate the capacity of the soil to absorb rainfall.
Soil depth	The depth of soil to bedrock or to an impermeable layer. Soil depth determines how deep roots, water and air can penetrate into a soil. This then influences how much water can infiltrate the soil, how much water can be held by the soil and how much soil plants roots can occupy.
Source control	The control of runoff at or near to its source.
Stakeholder	Person or organisation with a specific interest (commercial or professional) in a particular issue (political, regulatory, economic, financial, social, environmental etc).
Stormwater management tool	This tool uses several different methods to provide an initial assessment of water on site. The tool aims to establish: • storage requirements for drainage systems • the water management components most appropriate for each site.
Suitability evaluation tool	This tool helps assist in the optimum siting of water management technologies for new developments. The main output is a map to highlight where the water technologies chosen (eg SUDS) should be sited.
Surface water management plans	A framework for local partners with responsibility for surface water and drainage to work together to understand the causes of surface water flooding and agree the most cost effective way of managing that risk.
Sustainability criteria	Factors used to assess the range and combination that gives the most sustainable outcome. Typical criteria of sustainability are economical, environmental, social and technical.
Sustainability indicators	Indicators are sub categories of criteria that provide more specific detail on how to measure and assess the sustainability of a project, proposal or system. The indictors will depend on

	what is being assessed for sustainability and are likely to vary from site to site or project to project.
Sustainability principles	These are the wider umbrella terms that are used to establish how sustainability can be achieved.
Sustainable water cycle management	Incorporates different methods of water management to ensure water is conserved, reused and not wasted while still meeting the current needs of society, industry and the environment.
SUDS	Sustainable drainage system: a sequence of management practices and control structures designed to manage (drain, treat or/and contain) surface water at or near its generation point in a more sustainable fashion than other more conventional techniques.
Time-series rainfall	A continuous or discontinuous record of individual rainfall events generated artificially or selected real historical events that are representative of the rainfall in that area.
Tools	All software-related system components including but not restricted to spreadsheets, algorithms, decision support systems, simulation models, GIS components, libraries etc.
Ultra-low-flush toilet	WCs using less than two litres per flush.
Urban drainage	Pipe systems and other related structures to serve the removal of water from an urban environment.
Urban water optioneering tool	Decision support tool to help the selection of combinations of water-saving strategies and technologies and to support the delivery of integrated, sustainable water management for new developments.
Wastewater	Water used as part of a process that is not retained but discharged. This includes water from sinks, baths, showers, WC's, and water used in industrial and commercial processes.
Water conservation	Reduction in the amount of water used by limiting activities that use water or by ensuring that water is used more efficiently.
Water cycle study	A planning tool used by planning authorities and development organisations. The study identifies the capacity in water supply, wastewater infrastructure and water environment in growth areas to ensure that new developments can be supplied with the required water services it needs in a sustainable way.
Water efficiency	Reducing the amount of water to carry out a given task.
Water quality	The chemical, physical and biological characteristics of water with respect to its suitability for a particular purpose. The chemical and biological content of water that is usually compared to defined standards, many of which are set by national legislation or European Community directives and enforced by regulatory authorities in member states.
Water recycling	Reuse of water that originated from a potable tap.
Water-stressed	The existence of a high risk of failure in satisfying demand with the available water resources.

Water substitution	Replacing or reducing the amount of potable water used in a task.
Water table	The level of groundwater in soil and rock, below which the ground is saturated.

Abbreviations

ABI	Association of British Insurers
BATNEEC	Best available technologies not entailing excessive costs
BRE	British Research Establishment
CAMS	Catchment abstraction management strategies
CCW	Countryside Council for Wales
CEH	Centre for Ecology and Hydrology
CLG	Communities and Local Government
CSH	Code for Sustainable Homes
DALYs	Disability adjusted life years
Defra	Department for Environment, Food and Rural Affairs
DEX	Dunfermline Eastern Expansion
DRD	Department for Regional Development
EA	Environment Agency
EPSRC	Engineering & Physical Sciences Research Council
GIS	Geographic information system
GROW	Green roof water recycling system
GWR	Greywater recycling
HCA	Homes and Communities Agency
HIA	Health impact assessment
HFRB	Horizontal flow reed bed
ISSUES	Implementation strategies for sustainable urban environment systems
LANDF♦RM	Local Authority Network on Drainage and Flood Risk Management
LCA	Life cycle analysis
MATlab	Matrix Laboratory
MBR	membrane bioreactor
MCR	membrane chemical reactor
MTP	Market transfer programme
NAPI	Net antecedent precipitation index
NIAUR	Northern Ireland Authority for Utility Regulation
NIW	Northern Ireland Water
Ofwat	Office for Water Services Regulation
PAT	Project assessment tool
PR09	Price Review 2009
PPS	Planning Policy Statement

PPS1 (England)	Planning Policy Statement 1 *Delivering sustainable development*
PPS15 (NI)	Planning Policy Statement 15 *Planning and flood risk*
PPS25 (England)	Planning Policy Statement 25 *Development and flood risk*
RWH	Rainwater harvesting
SEPA	Scottish Environment Protection Agency
SET	Suitability evaluation tool
SST	Site screening tool
SUDS	Sustainable drainage systems
SUE	Sustainable urban environment
SWARD	Sustainable water industry asset resource decisions
SWCM	Sustainable water cycle management
TAN 15 (Wales)	Technical Advice Note 15 *Development and flood risk*
UKCIP	United Kingdom Climate Impact Programme
UKRHA	UK Rainwater Harvesting Association
UKWIR	UK Water Industry Research
ULFT	Ultra low flow toilet
UWOT	Urban water optioneering tool
VFRB	Vertical flow reed bed
WaND	Water and new developments
WCS	Water cycle studies
WDM	Water demand management
WFD	Water Framework Directive
WHO	World Health Organisation
WIC	Water Industry Commission
WLC	Whole-life cost
WRc	Water Research Centre
WSP	Water service provider

1 Introduction

This chapter:

- explains the water cycle in the context of sustainable water cycle management
- outlines the aims and objectives of this guidance
- describes the WaND project as the basis for the development of the guidance
- presents the context of the guidance
- highlights who should read this guidance and sections that are relevant for the various stakeholders.

This guidance addresses the challenges and solutions of sustainable water cycle management (SWCM) in new developments. It presents the concept of sustainability in relation to water cycle management and how it can be incorporated within decision making and delivery. It should help inform and influence decisions made during the master planning and delivery stages of water cycle management in new developments.

The guidance provides the evidence needed to achieve SWCM by presenting a better approach to planning for sustainable water management that to some extent is embedded in the planning process. It identifies a series of technological options able to improve the performance of new developments. Finally it considers engagement issues of stakeholders and health impacts surrounding water cycle management.

The guidance presents a selection of innovative tools that may assist in the planning of new developments and some have been developed for end users to aid this process. The appropriate deployment of the technologies and tools is highlighted. This guidance specifically supports the delivery of SWCM and raises awareness of the "carbon challenge" that is integral to the delivery of more sustainable solutions.

1.1 THE WATER CYCLE

The broader definition of the water cycle is the continuous movement of water through the environment above and below ground level. It includes the processes and systems that collect, store or transport water through evaporation and transpiration, condensation, and precipitation. The WaND project focuses more directly on the urban water cycle, from rainfall and drainage through to discharge and treatment.

The urban water cycle begins with the abstraction of water from rivers and aquifers to reservoir storage. The water is then processed through filtration and chlorination to a potable quality before being transported through an extensive pipework system to residential, commercial and industrial developments. After its use by humans, much of this water becomes wastewater and, along with some surface water runoff, is transported through a network of sewers to treatment plants, which, after treating it, discharge effluent into receiving waters such as rivers and the sea.

In most cases, the urban water cycle is managed on a large-scale centralised system incorporating water supply, wastewater and surface water treatment and distribution processes. Historically as part of the urbanisation process centralised networks have been used. The distribution elements of this system have existed in some civilisations for thousands of years, and the treatment aspects have endured for over 100 years. However, it is becoming increasingly apparent that the system is vulnerable where there are social factors such as high population density and high levels of water demand, and environmental factors such as water stress and flooding. Climate change and regional population growth are also exacerbating this problem. A decentralised approach to water management to supplement the centralised system is becoming increasing attractive, creating a society where water is also managed at a local scale to incorporate the three universal actions for sustainability: reduce, reuse and recycle.

1.2 AIM OF THIS GUIDANCE

The aim of this guidance is to repackage information from the water cycle management and new developments (WaND) research project and provide practical guidance to help stakeholders deliver SWCM for new developments. It is based upon evidence and important findings from the WaND research project (see Section 1.3). This document:

- sets the guidance in the context of current and proposed regulations and legislation
- identifies how sustainability can be incorporated within SWCM decision making
- outlines the main phases of a more sustainable approach to planning for water
- uses evidence from the WaND research project to:
  - review technologies that can be used to deliver SWCM
  - provide guidance on stakeholder engagement
  - introduce decision support tools for SWCM planning and design
  - present relevant case studies and good practice examples.

This document may also be used in conjunction with the Environment Agency publication on water cycle studies (WCS) (EA, 2009), which is principally aimed at local authority planners. The WCS guidance sets out the importance of water cycle management as part of the planning process and what the Agency expects when it is consulted by planning authorities on strategic planning documents through the local development framework or on a planning application. This guidance can be used to support the development of a water cycle strategy, by providing knowledge, information and tools for the different stages (scoping, outline and detailed studies) of a WCS.

1.3 THE WAND RESEARCH PROJECT

The water cycle management for new developments (WaND) research project was principally funded by the Engineering & Physical Sciences Research Council (EPSRC) between 2003 and 2007. It was led by the Centre for Water Systems at the University of Exeter and was completed by a consortium of academic institutions and industry collaborators (see Appendix A5 for a list of collaborators involved in the WaND project).

The aim of the project was to support the delivery of SWCM by providing tools and guidelines for project design, project implementation and management. The focus was on water supply, stormwater management and wastewater disposal in new developments. Further information on the project can be found in Appendix A1.

The WaND research project did not cover retrofitting existing buildings, nor did it look at the carbon cost of water management options so these are only briefly discussed within this guidance.

The scope of the WaND research project excluded flood risk management, which was separately covered by EPSRC's Flood Risk Management Research Consortium <www.floodrisk.org.uk>. This consortium was formulated to address the main issues in flood science and engineering by linking academic and industrial research partners.

1.4 CONTEXT OF THE DOCUMENT

This guidance is not intended to provide strategic planning on sustainable water management. Rather, once a site has been specified this guidance may be used to suggest tools, technologies and approaches to help those involved in the master planning and delivery stages of sustainable water management in developments.

The guidance explores water management at different spatial scales providing an overview of water management issues as well as measures that are available at a household scale. Detail is examined of some water efficiency measures and technologies to provide evidence for use on a smaller scale. This is justified as in some developments these small measures may be the only feasible measures.

Many planning policy statements (PPSs) emphasise sustainability and how to achieve it through strategic water cycle planning, including PPS1 and PPS 25. Water cycle studies provide a more sustainable solution to the challenges of growth, climate change and tightening standards, eg the Water Framework Directive (WFD). They are a new approach for delivering sustainability in the urban environment with respect to managing the urban water cycle. They identify the infrastructure required to support housing delivery, suggest development policies that should be applied by the planning authority to achieve the most sustainable solutions, and what management options should be promoted.

This guidance is intended to work with WCS once a site has been chosen by suggesting tools and an approach to support decision making at the master planning and delivery stage.

1.5 WHO SHOULD READ THIS GUIDANCE?

This document will be of interest to the following stakeholder groups:

- central government and regulators
- local government and planners
- developers, architects and consultants
- water service providers.

The guidance covers planning, delivery and management of the water cycle in new developments, from policy, through regional and local planning, to development and construction.

1.6 STRUCTURE OF THE GUIDANCE

Chapter	Summary	Target audience
1	• explains the water cycle in the context of SWCM • outlines the aims and objectives • describes the WaND project as the basis for the development of the guidance • presents the context of the guidance • highlights who should read this guidance and the sections relevant to the various stakeholders	All
2	• discusses the background to water management in the UK • sets out legislative, environmental and social pressures causing policy change as at September 2009 • discusses the policy, regulatory, research and responses to the various pressures for change.	Central government and regulators Local government and planners Water service providers

3	• outlines the challenges of water cycle management • presents the concept of sustainability by addressing: • what is sustainability? • how is sustainability linked to water cycle management? • how can sustainable water management be achieved? ◦ what is unsustainable development? ◦ what is sustainable development? ◦ sustainability and planning ◦ sustainability: principles, criteria, indicators • describes how scenarios can help plan for the future.	Central government and regulators Local government and planners
4	• outlines some of the technologies used in water supply management, stormwater management, wastewater management and water recycling/reuse. • sets out some of the technologies for water demand management and stormwater management, showing potential to help deliver sustainable water cycle management (SWCM) in new developments • summarises some of the available water-saving devices for household use including low-flush toilets, aerated showers, efficient washing machines etc • presents an overview of rainwater harvesting (RWH) systems to help water supply management and stormwater management • highlights some of the different treatment process associated with greywater recycling (GWR) and presents findings of research into five of the treatment processes available • illustrates the benefits of SUDS for stormwater management.	Developers, architects and consultants Water service providers
5	• identifies the factors that influence costs, benefits and risks to stakeholders (ie responsible bodies and end users including the general public), and the social issues and barriers to SWCM implementation • highlights opportunities to improve stakeholder engagement • identifies the main principles for achieving effective stakeholder engagement and presents ways to evaluate and frame key messages.	Central government and regulators Local government and planners
6	• presents the tools that where developed during the WaND research project to help manage the design and planning of urban stormwater management components and monitoring performance of sustainable drainage systems • discusses how the tools help manage sustainable water management, explains their functionality and how they can be developed.	Central government and regulators Developers, architects and consultants Water service providers
7	• provides important recommendations for sustainable water cycle management in new developments • summarises the messages for main stakeholders.	All

2 Water cycle management in context

This chapter:

- discusses the background to water management in the UK
- sets out legislative, environmental and social pressures causing policy change as at September 2009
- discusses the policy, regulatory, research and responses to the various pressures for change.

2.1 INTRODUCTION

This chapter provides context for the guidance on why water cycle management is important, especially in new developments. It examines the main pressures on the water cycle including environmental, social, economic and political factors.

Climate change, the Water Framework Directive and the Floods Directive have been key drivers in providing changes in policy, from a European to regional scale and will continue to provide further impetus in the future. The main outcomes in policy are highlighted in this chapter alongside the responses.

There have been several responses including government directives, research into sustainable water management and the delivery of sustainable water projects. The policy and legislation presented should be taken in context that these challenges are continuously changing as are the policy, regulatory and research responses.

2.2 BACKGROUND

The WaND research project was conceived in 2002 when government policy was to plan and build three million new households by 2020, while at the same time being committed to sustainable development. Since then, the UK Government has made several announcements on building large "sustainable communities" now known as "growth areas". It has also published the voluntary Code for Sustainable Homes (CSH) and water companies are expected to meet water efficiency targets while continuing to meet their existing targets on reducing leakage from the pipe network.

Future water (DEFRA, 2008a) was the Government's water strategy for England, setting a vision for water management now and in the future). In April 2008, the Communities and Local Government (CLG) announced their intention to promote several ecotowns as places of sustainable living and affordable housing.

In Northern Ireland the Department for Regional Development (DRD) has completed the transfer of responsibility for the delivery of water and sewerage services to a government-owned company called Northern Ireland Water (NIW). A new regulatory regime has been established with the formation of the Northern Ireland Authority for Utility Regulation (NIAUR) that will oversee NIW (as well as the gas and electricity utilities). The new arrangements set out the framework for the introduction of domestic charges for water and sewerage services, for the first time, as well as a commitment to an infrastructure investment programme.

In Scotland, the Scottish Government is working with Scottish Water, the Water Industry Commission (WIC) and the Scottish Environment Protection Agency (SEPA) to improve the quality of the water environment, deliver at least cost to the consumer while recognising the continuing need for infrastructure renewal (and investment). A key focus is on implementation of the Water Framework Directive.

In Wales, the Welsh Assembly Government has converted existing legislation covering England through Welsh Statutory Instruments, eg the Water Supply (Water Quality) Regulations 2001 (Amendment) Regulations 2007. Legislation and policy in Wales closely mirrors that in England but as regional government develops this may change and current Welsh planning guidance does not mirror the planning policy statements in England.

There has been growing international recognition of climate change and its potential future effects. The UK Government has responded to this by framing policies related to both mitigation through emissions reduction and adaptation to the likely effects. In England there have been several new planning policy statements, including PPS1 (PPS14 NI) and a supplement to PPS1. PPS 25 (SPP7 Scotland, PPS15 Northern Ireland and TAN 15 Wales) provide guidance on planning new developments with regard to flood risk and a better understanding of the need to maintain floodplains and allow them to operate naturally without the encroachment of development.

2.3 PRESSURES RESULTING IN POLICY CHANGE

Several factors have resulted in a review of policy on water and the way it is managed at all levels. These include environmental, social, economic and political pressures and are discussed in the following sub-sections.

2.3.1 Environmental pressures

Water quality

Water quality is an environmental pressure that has driven policy over the last few decades. The quality of the water in our rivers, lakes and estuaries can create amenity and recreational benefits and is often an indicator of how well we treat the environment. Water quality can enhance biodiversity and reduce energy costs because cleaner water needs less treatment to reach drinking water standard. Previously industrial processes and acute sources of pollution such as sewage treatment works and overflows have significantly affected water quality.

Over recent decades large investment has improved water quality however some issues still remain. This coupled with new pressures of population growth and climate change indicates that water quality is still an important environmental pressure.

Climate change and future climate scenarios

Climate change predictions are seriously exercising policy-makers. The Government published updated climate change predictions (UKCP09) in June 2009. They are used to help the UK plan for climate change, are based on the latest scientific understanding and contain information on observed and future climate change. The main findings of UKCP09 on how our climate might change are:

- all areas of the UK get warmer, and the warming is greater in the summer than in the winter

- there is little change to the precipitation (rain, snow, hail etc) that falls annually but it is likely that more of it will fall in the winter, with drier summers for much of the UK
- sea-level rise will be greater in the south of the UK than the north.

The UKCP09 provide details on the changes to the climate (ie more rainfall) but not the effects the climate has (ie where it might flood). This information will be used to feed into assessments of what the effects of a changing climate might be.

Rising temperatures, drier summers and changing precipitation events are likely to increase the strain on water resources. Although wetter winters could increase aquifer recharge, if the rainfall intensity increases, rapid surface water runoff could lead to less recharge and increased flooding. The predicted sea-level rises will also contribute to increased coastal flooding and saline intrusion into freshwater watercourses, with serious implications for low-lying coastal towns and agriculture.

Flooding

In the summer of 2007 severe rainfall events across large parts of the UK resulted in unprecedented rates of surface water runoff. Thirteen people lost their lives and over 55 000 homes and several thousand businesses were flooded leading to estimated insurance claims of £3.3bn. As a result, some home insurance premiums have increased significantly, some properties are "in danger" of being uninsurable, and the Association of British Insurers (ABI) has specifically asked the UK Government to reconsider any further development in floodplains. The Pitt Review (June 2008) suggested 92 recommendations for consideration. The Flood and Water Management Bill 2009 aims to give effect to the UK Government's response to that review.

Drought

It is often quoted that, in parts of south-east England, there is less water available per person than in several Middle Eastern countries. Seriously water-stressed areas are defined as those where current or future household demand is a high proportion of effective rain and/or, water abstraction is close to, or above, acceptable limits. A significant fraction of England is classified as being under serious water stress (Figure 2.1).

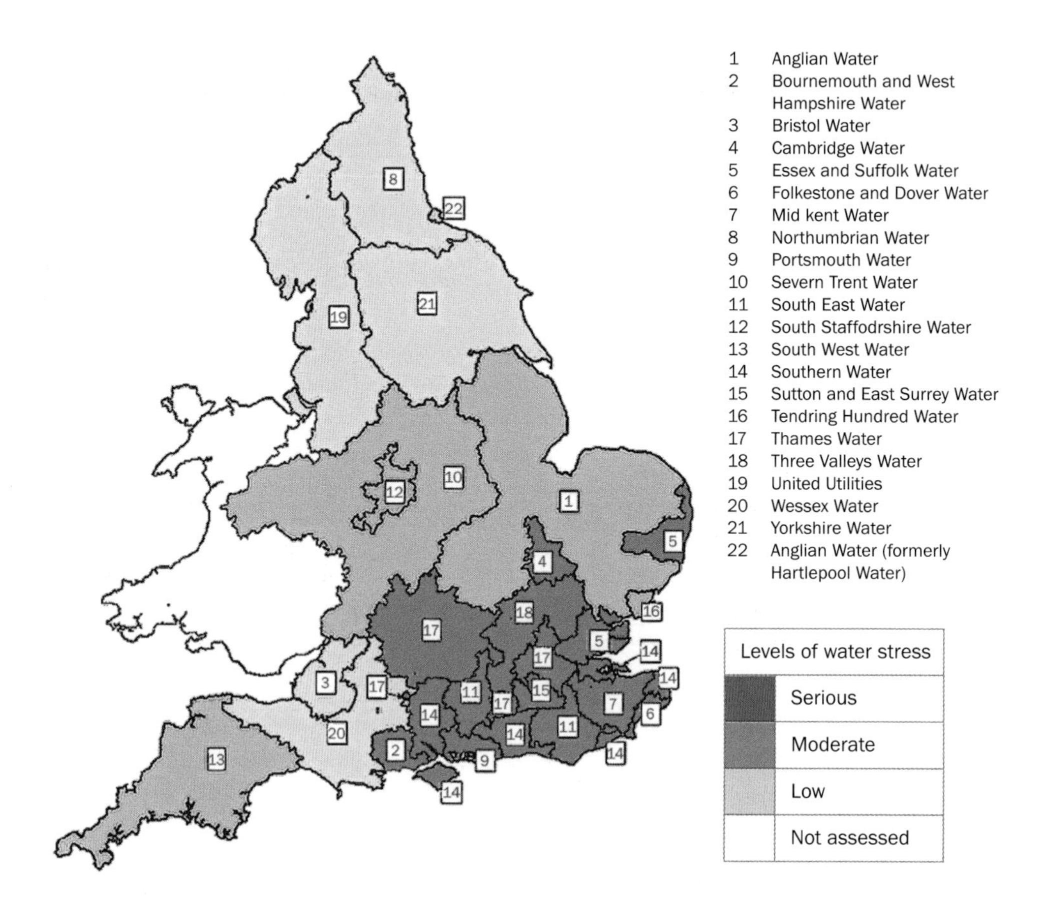

Figure 2.1 *Areas of relative water stress (Defra, 2008)*

The 2006 drought was a result of two consecutive winters of low rainfall that left groundwater aquifers at record low levels. A relatively short period of summer drought caused river levels to drop further, threatening the ability of water companies to abstract sufficient volumes or to replenish reservoirs. Those companies dependent on ground water sources found normally reliable boreholes close to drying up. To conserve supplies, widespread hosepipe bans and some drought orders were introduced. Rainfall in late summer 2006 averted the need for emergency drought orders. The water industry and its regulators came under close scrutiny from the media, the public and politicians and these droughts have influenced later environmental and planning policy measures.

Overall, increased climatic variability and an increase in extreme events will create significant difficulties both for the natural and the built environment. The environmental pressures clearly highlight that it is necessary for water service providers and consumers to significantly change their behaviour to fully contribute to the proposed reductions in carbon dioxide emissions and actively reduce water consumption.

2.3.2 Social and economic pressures

Increasing water consumption

Household water demand has been steadily increasing in the UK since the 1950s. This is due to population growth, changes in household size and changes in water use. There has been a very significant increase in the ownership of water-using appliances but also there have been changes in attitudes and behaviours to water use. This includes bathing (the increase in power showers), garden watering (the use of sprinklers and hoses) and even car washing (the use of jet power sprays).

It is estimated that the overall average water use in England is about 150 litres per person per day (150l/p/d) (Defra, 2009b). This equates to the equivalent of about 1 tonne of water per person per week. Domestic demand (partly because of the decline in heavy industrial process use) now accounts for 52 per cent of all public water supply use and based on Ofwat data for 2007 the total domestic demand is 7756 mega litres (millions of litres) per day (Ml/d). In addition, a further 5.85 per cent of all public water supply use (873Ml/d) is leakage on customer-owned pipework, where awareness of the problem and knowledge of the responsibility for getting the leaks repaired are not always very high.

Changes in housing stock

Projections suggest there is demand for at least 250 000 new dwellings per annum until at least 2020. This demand is largely attributed to factors such as increased rates of relationship breakdown: the need for single-occupancy dwellings for young and old alike, financial pressures creating demand for smaller affordable housing units and increased demand from immigration. All these trends are shifting in the direction of a sustained increase in smaller-occupancy housing and it is estimated that between 2002 and 2021 one-person households will increase from 6.1m to 8.5m.

The demand for new housing is often higher in regions that are already water stressed (Figure 2.1). By 2016, Thames Water expect to supply 700 000 more people in London, the equivalent of a city the size of Leeds moving in to the capital. This will cause further strain on water resources in one of the driest areas of the country.

Communities and Local Government (CLG) are responsible for determining new "growth points" throughout England. In October 2006 45 towns and cities were confirmed as having the potential to deliver up to 100 000 new homes (over and above those previously planned) in the next 10 years. Alongside this CLG are also looking to promote up to 10 eco-towns (up to 10 000 homes) as exemplar sustainable developments.

2.4 POLICY PRESSURES

2.4.1 European Union Directives

EU Directives are the most influential driver for new environmental policy in the UK and they set out broad requirements that all member states need to follow. Four directives have particular influence and are important to consider: the Water Framework Directive, the Urban Wastewater Treatment Directive, the Floods Directive and the Bathing Waters Directive.

The Water Framework Directive (WFD) (2000)

The Water Framework Directive (WFD) was enacted to improve and integrate the way water bodies are managed throughout Europe. The Directive came into force at the end of 2000 and was transposed into UK law in 2003. Much of the implementation work is undertaken by the Environment Agency, the Countryside Council for Wales (CCW) and Scottish Environment Protection Agency (SEPA) through river basin management plans. The Directive's main goal is for member states to reach good chemical and ecological status in inland rivers and coastal waters by 2015. Other main aims are to:

1. Improve the status and prevent further deterioration of aquatic ecosystems and associated wetlands that depend on those aquatic ecosystems.

2. Promote the sustainable use of water.
3. Reduce pollution of water, especially by "priority" and "priority hazardous" substances.
4. Ensure the progressive reduction of groundwater pollution.

The Directive affects all areas of water use and is the single biggest influence on change in UK water policy. Two areas particularly relevant are:

1. The tightening of licences governing water company abstractions (especially in dry months) to maintain good river water quality status. This has further emphasised the need for sustainable demand management.
2. Tighter control on both point source and diffuse pollution make a stronger case for managing urban drainage sustainably and removing pollutants before they reach rivers.

The Urban Wastewater Treatment Directive (1991)

This Directive is designed to protect the environment from the adverse effects of urban wastewater discharges and those from certain industrial sectors. It concerns the collection, treatment and discharge of domestic wastewater, mixed wastewater and wastewater from some industrial sectors. This Directive affects planning (designating certain areas), regulation (authorisation of discharges and treatment requirements), monitoring (analytical methods, sampling frequency and parameter definitions) and reporting requirements (disposal and reuse of urban wastewater and sewage sludge).

The Floods Directive (2007)

The European Directive on the Assessment and Management of Flood Risks is designed to help member states prevent and limit floods and their damaging effects on human health, the environment, infrastructure and property. The Directive came into force in November 2007 and member states have two years in which to transpose it into domestic law. The Directive requires member states to prepare the following assessments for the European Commission:

- preliminary flood risk assessments to identify areas that are at potentially significant flood risk, by 2011
- flood hazard maps (showing the likelihood and flow of the potential flooding) and flood risk maps (showing the effects), by 2013
- flood risk management plans (showing measures to decrease the likelihood or effect of flooding), by 2015
- from 2015, to update every six years to take into account the effect of climate change.

The assessment process must be aligned with the environmental objectives of the Water Framework Directive.

The Bathing Waters Directive (1976 and 2006)

The main objective of the European Commissions Bathing Waters Directive 1976 was to protect public health and the environment from faecal pollution to bathing waters. The Directive was revised in 2006 to reinforce the protection of public health, to reinforce the sampling and reporting requirements and to align this directive with the Water

Framework Directive (as bathing waters are protected areas under WFD). This directive had a very significant effect on treatment processes for coastal communities and the cessation of sewage disposal at sea.

2.5 RESPONSES

The responses to the pressures outlined in Section 2.3 have been subdivided into three categories: policy and regulatory, research, and implementation.

2.5.1 Policy and regulatory responses

UK Climate Change Programme (2006) and Climate Change Act (2008)

The Stern Review (Stern, 2005) on the economics of climate change was the first annual report to parliament on climate change in England in July 2007. Following this report the UK Government signalled a need for a clear strategic approach to climate change and has produced a legislative framework in the Climate Change Act 2008.

The main provisions of this Act include:

- transitioning towards a low carbon economy in the UK by improving carbon management
- 80 per cent reduction in green house gas emission by 2050 and 26 per cent reduction of CO_2 emissions by 2020
- publishing of five year carbon budgets by the government
- creating a climate change committee to advise on the level of carbon budgets to be set
- placing a duty on government to assess the effects of climate change and the risks facing the UK
- providing powers to establish carbon trading schemes.

Water cycle management will have a direct effect because the whole water cycle is now under close scrutiny of its carbon output and this will require significant behavioural changes from regulators, water service providers and all water consumers.

In Wales provisions embedded in the UK Bill require actions of Welsh ministers in terms of reporting and producing guidance and giving support to public authorities on climate change adaptation. Northern Ireland received a report on preparing for a changing climate in January 2007 and the Department of Environment (through the NI Climate Change Impacts Partnership) is now looking at how UK legislation translates into local adaptation and resilience measures as well as targets for emissions reduction. In Scotland, the Climate Change (Scotland) Act 2009 sets a statutory target to reduce emissions by 80 per cent by 2050 and requires the setting of annual targets for the reduction of greenhouse gas emissions. It also imposes a duty on Scottish Water to promote water conservation and water-use efficiency.

Future water – the Government's water strategy for England (2008)

This is the first statement of water strategy since 2001. A central theme in the strategy is reducing demand for water. The vision is to see domestic water consumption reduced to 130 litres per person per day (l/p/d) by 2030. Defra see the reductions in demand being achieved through high standards of water efficiency in new homes, retro-fitting water-saving devices in the existing housing stock and further reductions in leakage (both from the pipe network but, significantly, from householder-owned pipework).

Future water opens the debate on "water neutrality" in planning and designing new housing such that the combined water demand for both the new and existing users is not increased when new development is permitted. This suggests radical new approaches to adopt retrofitting water-saving devices, rainwater harvesting, the design of lower water-using appliances, and a substantial change in consumer behaviour. The Government expects planning authorities to work closely with water companies and regulators on the timing and numbers of new homes to be built.

The strategy highlights the scope for capturing rainfall on the ground and on domestic and residential properties, before it runs into rivers and drainage systems. It connects ecologically sensitive planning to whole river catchments by increasing the scope for floodwater retention in managed soakaways and wetlands; to planning the site and design of residential and commercial property so that permeability is improved, impermeable surfaces are inhibited, and rainwater harvesting is encouraged.

UK Government also amended the Town and Country Planning General (Permitted Development) Order 1995 so that permission is needed for paving over front gardens unless permeable surfacing materials are used. In areas where risk from surface water drainage is significant, Defra advocates the preparation of surface water management plans to be agreed by all local stakeholders with drainage responsibilities, clarify responsibilities and help manage the risks. Defra suggests that water companies need to proactively support these plans through expertise, data and overland-flow models. It also suggests that the Environment Agency should provide input and review the assessments and where necessary, object to local plans and planning applications on drainage grounds. The Government is also considering whether funding for surface water drainage should be changed to better reflect the "polluter pays" principle. There is consultation on how to ensure that surface water management plans are properly considered for all development and investment planning. For house builders and homeowners this means measures to decrease the amount of water running in to drains by using sustainable drainage systems (SUDS).

HM Treasury: Public Service Agreements (PSA), 1–7 and 27–30

The Public Service Agreement (PSA) sets out the main priority outcomes the UK Government wants to achieve in the next spending period (2008–2011). PSA 27 sets out to lead the global effort to avoid dangerous climate change, and includes several measurement annexes. Indicator 2 measures the proportion of areas with sustainable abstraction of water as defined by the catchment abstraction management strategies (CAMS) baseline data available in March 2008. CAMS are the Environment Agency strategies developed in consultation with local people, designed to help the EA's licensing of abstractions. Sustainable abstraction is measured as the proportion of the total number of Environment Agency CAMS that are not either "over-abstracted" or "over-licensed". This will put significant pressure on water companies to push the reduction of customer demand, particularly in drier months.

Development and flood risk policy statements

PPS 25, PPS 15 (NI), SPP7 (Scotland) and TAN 15 (Wales) outline the way local authority planners and developers are required to approach the management of flood risk to ensure it is taken into account at all stages in the planning process. This is to avoid inappropriate development in areas at risk of flooding or, if the development is absolutely necessary, to make it safe, without increasing flood risk elsewhere, and, where possible, reducing flood risk overall. These policy statements also advocate green infrastructure for flood storage and conveyance and sustainable drainage systems (SUDS).

The Town and Country Planning (Development Procedures) Order, as amended, established the Environment Agency as a statutory consultant for development in flood risk areas. The Order requires local planning authorities minded to approve applications in the face of sustained objection by the Environment Agency to refer those applications to the government office for the region to determine whether they should be called in for ministerial decision. This should enable better integration of risk assessments in planning decisions, particularly in the context of the strategic flood risk assessments recommended in PPS 25.

A practice guide was published in June 2008 as a complementary document for PPS 25 and explains how to implement policies in practice (CLG, 2008). It is aimed at regional and local planning authorities and draws on existing good practice of case studies to highlight varying circumstances.

Code for Sustainable Homes (England and Northern Ireland)

The Code for Sustainable Homes, published as a voluntary standard in April 2007, measures the sustainability of a new home. In parallel *Building a greener future – towards zero carbon development* (CLG, 2006), reviewed whether the carbon rating against the Code should be mandatory. In Northern Ireland all public housing requires a Code rating.

The Code sets standards for a wide range of sustainability indicators that are not mandatory in the current Building Regulations but are critical to limiting the environmental effect of housing. With specific reference to water, all social housing funded through the Homes and Communities Agency (HCA) need to be built to level 3 of the Code and achieve overall household use that equates to 105 l/p/d from April 2008. This figure is significantly below the *Future water* vision for 2030 that all households will reduce per capita consumption to 130 l/p/d.

The CSH technical guide is updated every six months and a dwelling needs to meet the requirements of the Code when registered with a Code assessor.

In Scotland and Wales, the Code does not apply and the ecohomes and ecohomes XB measures (developed by British Research Establishment) will continue to be used to assess refurbished and existing homes.

Water calculator

The CSH monitors water efficiency in buildings through the water calculator. The water calculator is used as a benchmark assessment of the typical consumption of a specific water fitting. It calculates the contribution that a fitting will make to the whole house and the water consumption measured in litres per person per day. The contribution of each of the fittings is calculated and the house is then given an overall water consumption per person that equates to a Code rating.

Defra's adaptation policy framework

The Climate Change Act 2008 requires the development of a programme to adapt to the effects of climate change. This programme has two phases. It is now in the first phase, 2008-2011, and includes:

- creating a better understanding of climate change
- creating a greater awareness
- measuring success and ensuring steps are taken to ensure the effective delivery
- working across all scales of government to embed adaptation into government policies, programme and systems.

The second phase is the implementation of the statutory national adaptation programme, which will be developed once the groundwork of the first stage is completed. The policy emphasises that the adaptation needs to occur at all scales from government, local authorities, businesses, charities and individuals if the negative effects of climate change are going to be reduced. It is important that changes to the decision making process are needed now because, without early and strong mitigation, the physical limits and cost implications will grow rapidly.

Building Regulations

Communities and Local Government (CLG) is making changes to Part G of the Building Regulations (sanitation, hot water safety and water efficiency) to set water efficiency targets for new dwellings at 125 l/p/d. The Regulations will provide a water consumption calculator, which will be a simplified version of the one used in the CSH. These Regulations cover England and Wales.

In Scotland in 2001, Building Standards (Scotland) Regulations 1990 amended Part M (Regulations 24, 25 and 25A) covering drainage and sanitary facilities. The amended Regulation to 24 covers household water disposal and drainage of surface water form buildings and paved surfaces within the curtilage of the building. Regulation 25 requires the provision of adequate and suitable sanitation facilities in all buildings.

In Northern Ireland the Department of Finance and Personnel is responsible for the implementation of policy and legislation relating to Building Regulations. Part N deals with drainage and Part P deals with sanitary appliances and unvented hot water storage.

Water metering and tariff structures

The current system of charging for water, based on the rateable value of a property is increasingly being seen by policy-makers (and many customers) in England as indefensible, particularly in water-stressed regions. Now about 30 per cent of all customers have a water meter, which means that for 70 per cent their water bill bears no direct relationship to the actual water use. Metering is increasingly seen as the fair way to pay for water as customers pay for what they use and this introduces a financial incentive to save water by using less. Metering can stimulate water efficiency and the evidence base suggests that fitting a water meter reduces household consumption by about 10 per cent. The challenge is whether or not the regulatory process can (and will) encourage the development of tariff structures that will result in demand suppression and reductions in water consumption.

The regulatory framework was adjusted in 2007 to allow water companies in water-stressed regions in England to consider compulsory metering. The companies involved have stated their commitment to increased metering in their recently published strategic direction statements, required of them by Ofwat (2007). Further evidence of a wholesale intention to increase metering may become clearer when all the water companies have published their water resource management plans and when the PR09 price review process has been completed.

In Northern Ireland the whole process of "water reform" is stalled while acceptance for billing is decided upon. The proposed mechanism is to use house value to define the size of the bill. In Scotland a process of harmonising charges based on council tax bands was undertaken in 2004–2005 and in Wales there is a similar situation although collection is in some cases made by the local authority on behalf of the water service provider. So only in England will water bills increasingly reflect actual use.

Water Supply (Water Fittings) Regulations 1999

Since 1 July 1999 in England and Wales and 4 April 2000 in Scotland and Northern Ireland, these Regulations have replaced the previous UK water byelaws. The Regulations were made for England and Wales under Section 74 of the Water Industry Act 1991 to prevent waste, misuse, undue consumption (as well as contamination or erroneous measurement) of drinking water. There was a further Statutory Instrument in 2005 (No 3077) but the UK Government set up a new review of these Regulations in 2008 to consider maximum water use of toilets, urinals and washing machines, and also to consider advances in technical standards and water conservation and the case for setting new performance standards for vital water fittings (for example, shower heads).

Statutory social and environmental guidance to Ofwat (Defra, 2008c) England and Wales

The guidance sets out how the UK Government expects Ofwat to contribute to main areas of social and environmental policy set within the context of sustainable development. Although Ofwat is not bound to act on the guidance it will have to demonstrate in its Annual Report how it has contributed to the achievement of sustainable development and it is expected to work with organisations such as the Environment Agency and Natural England and consider environmental outcomes in their broadest sense.

This marks a clear shift in what the Government expects from the water industry's economic regulator in England and Wales and indicates that sustainable development is expected to reach across the whole of government. In Northern Ireland there is no formal periodic review process and the relationship between Northern Ireland Water and the utilities regulator is in its infancy. In Scotland the relationship is between Scottish Water and the Water Industry Commissioner (WIC) and it is unclear as to what extent, if any, the expectations placed on Ofwat will also be placed on the WIC.

Water company strategic direction statements (England and Wales)

Following a letter from Ofwat to all water companies in 2007, strategic direction statements have been drawn up by each company after consulting customers and stakeholders on a range of issues. The statements set out the long-term priorities for the company, over a 25 year period. They give each company the opportunity to address the following:

- plans to deliver for consumers and the environment in the long-term (ie at least 25 years ahead) including clean and safe drinking water and reference to the water sector plan
- the effect this has on, among other things, the management and stewardship of assets, and innovation
- the approach to issues such as climate change and sustainability
- a clear understanding of customers' priorities
- the objectives of a long-term charging strategy, the plans to achieve this and the implications for customer's bills
- the major risks to the business and how these will be managed
- the financing of the strategy.

Ofwat's consultation on water efficiency targets

Ofwat has proposed a programme for water efficiency targets for each water company. The targets will initially be in place for five years 2010–2011 to 2014–2015 and a review of these targets will take place during this period.

The base requirements for each company will be to:

- reduce water consumption per household to a target based on the number of properties served by the company
- to continue to provide information to consumers on how to be more water efficient
- contribute to improving the evidence base for water efficiency.

The target would be based on a reduction of one litre per household each year. Across England and Wales this reduction would represent a 40 per cent increase in the average amount of water saved through water efficiency activity.

Water companies would be expected to pursue further water efficiency targets especially through innovative water efficiency solutions. These would be formed as part of a sustainable economic approach to balancing supply and demand.

Water for people and the environment – water resources strategy for England and Wales (EA, 2009b)

The new EA water resources strategy sets the high-level strategy for water resources in England and Wales, details the principles for managing water resources and outlines the expectations of other stakeholders. During preparation of the strategy the EA have published several working papers for consultation:

- water resources in south-east England
- carbon and energy use in water resources
- valuing water resources
- water resources in 2050
- governance and legislation for water resources
- water resources strategy forecasts.

The Environment Agency approach to water companies resource management plans

Water companies have to provide a 25 year plan to the EA to show how they intend to supply water over that period taking into account population change, climate change and the protection of the environment from over abstraction. The provision of these plans allows the EA to have greater control over the methods used by the water companies to supply their customers, which includes creating a greater balance between new resources and making better use of the resources that are already available. The EA has set out guidelines for water companies including the need to:

- take further steps to reduce leakage
- consider climate change and reduce any damage to the environment due to extraction
- consider and, where possible, reduce the carbon footprint of current and future operations
- ensure forecast demands are inline with the Government's commitment to reduced water and carbon in new homes
- encourage metering in households whenever possible.

The Water Savings Group (England)

The Water Saving Group was established in 2005 and brings together water industry stakeholders to share knowledge and promote the efficient use of water in households in England. It seeks to identify ways of raising awareness about water efficiency, as well as identifying collaborative work to improve the evidence base and contribute to demand management and the long-term sustainability of water supplies. The group aims to find ways to reduce the current level of per capita consumption in households.

The Flood and Water Management Bill 2009

The Flood and Water Management Bill was laid before Parliament in November 2009. It aims to create a healthier environment, better services and greater protection for people and their communities and business. The Bill will affect SWCM in new developments by:

- delivering improved security, service and sustainability of water for people and their communities
- clarify who is responsible for managing all sources of flood risk
- protecting vital water supplies by enabling water companies to control more non-vital uses of water during droughts
- encouraging more sustainable forms of drainage in new developments
- making it easier to resolve misconnections to sewers.

One of the main outcomes that will affect water cycle management in new developments is the shift from building flood defences to flood risk management. Water on site will need greater consideration especially as the Bill will end the automatic right for a new development to connect to the sewer system for surface drainage. Instead there will be the requirement to use SUDS to manage surface water. Further to this it is proposed that SUDS will also be adopted and maintained by local authorities.

Legislation on hosepipe bans will also be updated to enable water companies to conserve water from an earlier stage of a drought to ensure that supplies continue to be available during drought periods.

2.5.2 Research responses

This section identifies what research there has been in the area of sustainable water cycle management and where to find important information.

Sustainable urban environment

The most ambitious and comprehensive research in the sustainable development field has been sponsored by the Engineering and Physical Sciences Research Council (EPSRC) under the sustainable urban environment (SUE) programme. Several large multi-disciplinary consortia have been funded, involving more than 30 UK universities and over 120 project partners including local authorities, large and small companies, town planners and charities. The overall aims of the programme are to:

1. Develop and promote a strategic research agenda to address sustainability in the urban environment for the 21st century and beyond.
2. Strengthen the capability of the UK research base in sustainability issues within the urban environment.
3. Engage end users of research in industry, commerce, and the public and service sector.

The dissemination from the programme is being co-ordinated by the Implementing Strategies for Sustainable Urban Environment Systems (ISSUES) project. WaND is the consortium covering aspects related to water and its future management.

CIRIA

CIRIA is a not-for-profit research organisations encouraging good practice within the built environment and construction sectors. It has a long track record of applied research, guidance development and dissemination. Relevant research guidance includes:

C539 *Rainwater and greywater use in buildings. Best practice guidance* (Leggett *et al*, 2001a)

PR080 *Rainwater and greywater use in buildings. Decision-making for water conservation* (Leggett *et al*, 2001b)

C625 *Model agreements for sustainable water management systems. Model agreements for SUDS* (Shaffer *et al*, 2004)

C626 *Model agreements for sustainable water management systems. Model agreement for rainwater and greywater use systems* (Shaffer *et al*, 2004)

C630 *Sustainable water management in land use planning* (Samuels *et al*, 2005)

C635 *Designing for exceedance in urban drainage – good practice* (Digman *et al*, 2006)

C697 *The SUDS manual* (Woods Ballard *et al*, 2007)

C680 *Structural design of modular geocellular drainage tanks* (Wilson, 2008)

UKWIR

The UK water industry "one-voice" research is managed by UK Water Industry Research (UKWIR). Their main contribution to date, to water cycle management has been in the area of water efficiency guidance for water companies:

05/CL/04/6 (2003a) *Effect of climate change on river flows and groundwater recharge*

03/WR/25/1 (2003b) *Quantification of the savings, costs and benefits of water efficiency*

05/WR/29/1 (2005) *Framework for developing water reuse criteria with reference to drinking water supplies*

06/WR/25/2 (2006a) *Sustainability of water efficiency*

07/WR/25/3 (2007) *A framework for valuing the options for managing water demand*

Waterwise

Waterwise is a UK NGO focused on decreasing water consumption in the UK by 2010 and building the evidence base for greater water efficiency. Presently, it is involved in several large scale water efficiency projects throughout the UK, often in partnership with the water industry to develop the evidence base to save water. They are increasingly working with the Energy Saving Trust on the carbon effect of water cycle management.

European projects

The main ongoing European project is SWITCH funded by the European Commission and adopted by a cross-disciplinary team of 33 partners and 13 cities in 15 countries around the world. SWITCH aims to produce a change in urban water management, away from existing ad hoc solutions towards a more coherent and integrated approach. The vision of SWITCH is for sustainable integrated urban water management in the "city of the future".

2.5.3 Implementation responses

This section briefly identifies where there have already been significant responses to the policies in sustainable development.

Water cycle studies in England and Wales (EA)

WCS are principally aimed at providing strategic planners of local authorities a non-prescriptive methodology to assess the infrastructure and environmental capacity for sustainable water cycle management within new large-scale housing developments. Where a lack of capacity exists, WCS act to help planners make decisions on how to meet this deficit with sustainable solutions (see Chapter 3 for more information about WCS).

The EA will only be seeking WCS as a "mandatory approach" for eco-towns and new growth points. The WCS guidance will have the status of good guidance and will not be formal EA policy.

CLG eco-towns – England and Wales

CLG have designated locations for eco-towns (each designed as self-contained communities of up to 10 000 homes). The policy and guidance encourages the planning process to fully recognise the need for water infrastructure capacity in these areas and the Environment Agency want to see water cycle studies as a mandatory part of the process for all new growth point and eco-town designations. Alongside this it will also be important that the water company water resource management plans provide a clear picture of overall system capacity and system constraints and that these inform the planning process at an early stage. There is the opportunity to deliver "joined-up" government if the planning process is able to recognise all the issues that contribute to making eco-towns sustainable developments.

The locations assessed as having potential for an eco-town were identified in the publication of the Planning Policy Statement: *Ecotowns and supporting documents* (CLG, 2009). These locations are identified in Figure 2.2. The document also sets out the highest ever standards for green living.

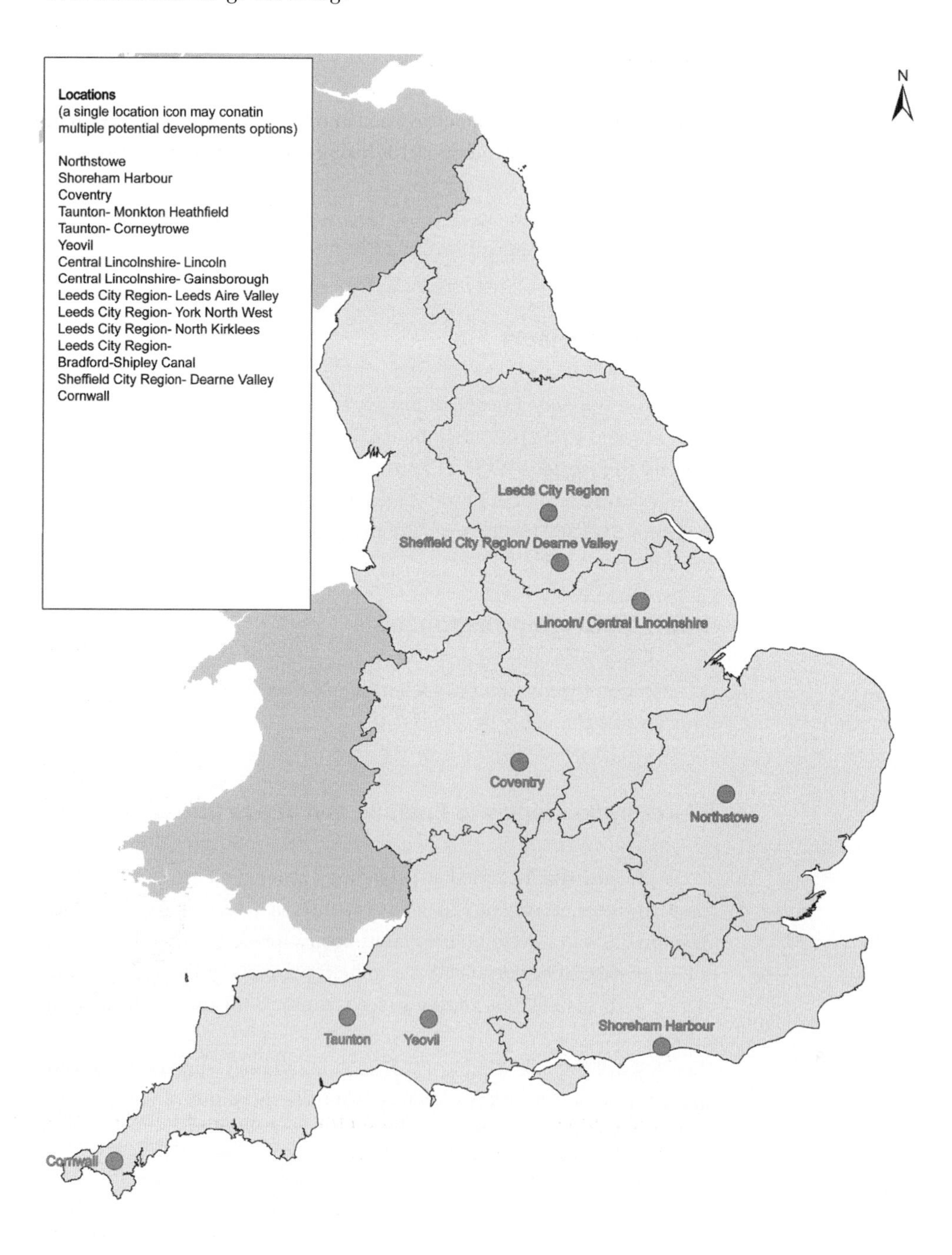

Figure 2.2 *Possible eco-town locations (CLG, 2009)*

Water neutrality

Water neutrality suggests that a new development should not lead to an overall increase in the demand for water. The water requirement of the new development would be met through reducing water demand in the area surrounding the development by retrofitting existing homes and other buildings with more efficient devices and appliances.

The Environment Agency is exploring how the concept of water neutrality can be used to encourage water efficiency. The concept differs from normal water resource planning because it is an aspiration being explored through exemplar developments. CLG, Defra and the Environment Agency facilitated the investigation of achieving water neutrality in the Thames Gateway by commissioning the study *Towards water neutrality in the Thames Gateway* (EA, 2007a). The study found that by delivering an ambitious plan of demand management that the demand for water in the area could remain the same in 2016 as in 2005 even with the introduction of 165 000 new homes.

A scenario based approach was used and it was found that water neutrality can theoretically be achieved by 2016 through a combination of measures including:

1. Building new homes to higher standards of water efficiency.
2. Improving the water efficiency of existing homes through retrofitting of water-saving appliances.
3. Metering of new and existing homes.
4. Introducing variable tariffs.
5. Improving the water efficiency of non-households.

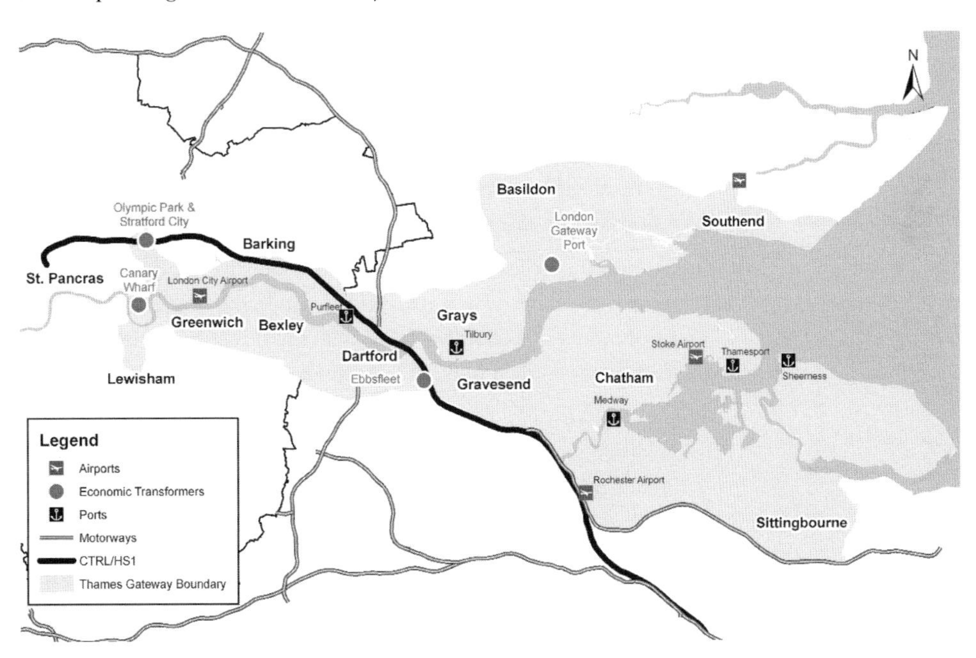

Figure 2.3 *Thames Gateway (source Homes and Communities Agency)*

The Environment Agency has led on this concept and is looking at opportunities to build on the Thames Gateway study and investigate how to define and deliver water neutrality.

2.6 CARBON REDUCTION AND THE WATER CYCLE

Carbon reduction is vital for SWCM. The Climate Change Act 2008 highlights the importance of reducing CO_2 emissions and to date energy and water have often been treated as separate entities but their links are becoming more evident. Current water management practices use large amounts of energy in the treatment and pumping of drinking and wastewater.

The water industry is responsible for high energy consumption which account for around one per cent of the total UK CO_2 emissions (Waterwise, 2009). However the main link between energy consumption and water is household water use. The use of water in our homes for activities such as personal use, household washing, cooking and cleaning accounts for five per cent of the total CO_2 emissions in the UK (Defra, 2008). The majority of this energy is used in the heating of hot water and accounts for around 25 per cent of the average household energy bill. This highlights that there are strong links between energy and water and that water efficiency will help reduce CO_2 emissions.

The adoption of a demand-management approach should have the potential to save water and energy and also reduce the carbon footprint throughout the water system. This approach includes encouraging the use of water-efficient appliances and behavioural changes such as reducing showering time and using full loads in washing machines and dishwashers. These simple measures by individuals will result in significant reductions in greenhouse gas emissions while also lowering energy and water bills. Further to this the Environment Agency's work on energy and water demonstrates that building new resources schemes such as desalination and new reservoirs will increase greenhouse gas emissions whereas helping customers reduce their water use will reduce emissions.

2.7 CONCLUSIONS

- this chapter demonstrates the various pressures on the water cycle resulting from environmental issues and policy initiatives, as well as social and demographic change
- the responses to date have been equally varied both at the policy level and in terms of regulation and the research outputs
- although the responses are broadly welcomed, it is clear that government measures are not yet sufficiently joined-up. For instance, there is no clear agreement of what per capita consumption should be or consistency on how the water cycle can be joined up
- to achieve the necessary reductions in water use, traditional approaches to water management have to be replaced by sustainable water cycle management in new developments as well as the wholesale introduction of adaptation measures (of which retro-fitting is only a part) in the existing housing stock
- the water neutrality concept needs to be proven at a large scale and it is hoped that developments at Thames Gateway, Ashford and the new eco-towns can provide further evidence on how neutrality can be delivered
- water cycle studies need to become a routine part of the planning process so that pressure to build on floodplains and in water-stressed regions can be very carefully evaluated in terms of the capacity of existing infrastructure and the environment
- regulation of the water service providers needs to provide real incentives if they are to fulfil their duty to deliver improved water efficiency and real savings in carbon dioxide emission from source to tap
- sustainable development has historically focused on building materials, energy efficiency and carbon costs and much more attention needs to be given to SWCM and the comprehensive benefits this may deliver
- understanding the interaction between carbon emissions and water management are important in this regard but it is also essential that the general public understand the value of water and the need to change behaviour in terms of how this valuable resource is used on a daily basis.

3 Planning for water

This chapter:

- outlines the challenges of water cycle management
- presents the concept of sustainability by addressing:
- what is sustainability?
- how is sustainability linked to water cycle management?
- how can sustainable water management be achieved?
    - what is unsustainable development?
    - what is sustainable development?
    - sustainability and planning
    - sustainability: principles, criteria, indicators
- describes how scenarios can help plan for the future.

3.1 INTRODUCTION

Effective integration of water management into the planning system is fundamental to managing resources sustainably, and to the overall national goals of sustainable development. This chapter explores the concept of sustainable development, what it means for urban water cycle management and how it can be put into practice through the use of principles, criteria and indicators.

Recognising what sustainability is and how it relates to the water cycle is the first step in understanding how to achieve it. This chapter seeks to explain sustainability, and how it should be determined by site-specific and regional factors rather than a set of generic criteria that can be applied in every situation. Different outcomes are inevitable so are demonstrated in the case studies on sustainable water cycle management (SWCM) in Section 3.5.2.

3.2 CHALLENGES

Challenges for water cycle management are varied and include alignment with government policy (economic, environmental and social), compliance with directives and regulation, harmonisation with spatial development, anticipation of future developments and vigilance for global change, eg changes in climate, demographics, human behaviour (Hurley *et al*, 2007a).

In general, water infrastructure delivery, replacements and upgrades are expensive, long-term decisions that need to be integrated within relatively short-term budgets and planning horizons. Most new infrastructure affects a range of stakeholders, all with their own drivers, concerns and constraints, which when integrated into the decision making process, can increase complexity.

Most importantly, the long-term trends suggest that demand for homes in England is increasing at a faster pace than new homes are being built and communities are increasingly urbanised (over 90 per cent of the population lives in urban areas covering just eight per cent of the land area). This has a significant effect on the management of the water cycle.

Government objectives for sustainability are delivered through several policy documents (see Chapter 2). In particular, PPS1 and the Government's sustainable communities plan (ODPM, 2003) set out a strategy for accommodating new housing requirements and delivering affordable housing and job creation as well as protection of the environment.

The sustainable communities plan aims to avoid urban sprawl by building at a higher density and by redeveloping brownfield sites. Increased densities and reduced green space may decrease some options for sustainable surface water management and development on contaminated land presents further challenges for water management.

With these issues, decision making for sustainability is not straightforward, because the overall objective is not easy to define.

3.3 WHAT IS SUSTAINABILITY?

Sustainability is a broad concept that needs to be defined, but there is general agreement that the main components include society, the environment and the economy. In the Brundtland Report (Woods-Ballard *et al*, 1987), the definition of sustainability "...to serve the needs of today without compromising the needs of tomorrow's generations" is a widely accepted definition and has supported a more inclusive approach to decision making.

Others state that sustainability is the move away from the relationship between society and nature being a vulnerable, destructive system, to it being resilient, constructive and economically productive. PPS1 highlights that the main feature in sustainable development is ensuring a better quality of life for everyone now and in future generations.

Historically we have built and designed our urban environment to meet the needs and capacity of the time. Some of our infrastructure is very old and was built a hundred years ago and has reached its capacity. As a result, much of our existing urban building stock and infrastructure may leave us vulnerable to the effects of social, environmental and economic change. This includes our ability to cope with fewer or more expensive resources, manage extreme weather events, adapt to climate change, deal with urban population growth or meet environmental targets and legislation.

The UK Government has proposed a sustainable development strategy, from its specific stakeholder perspective, with the context based on the Brundtland Report's definition. It includes five underpinning principles of sustainability (see Figure 3.1), which were agreed by the UK Government (including the Northern Ireland Assembly, the Welsh Assembly and the Scottish Executive) and are reflected in their strategies.

The aims of the government's strategy to achieve sustainability are:

- social progression that recognises the needs of everyone
- effective protection of the environment
- the prudent use of natural resources
- the maintenance of high and stable levels of economic growth and employment.

For a further discussion on sustainability and current academic ideas refer to Appendix A2.

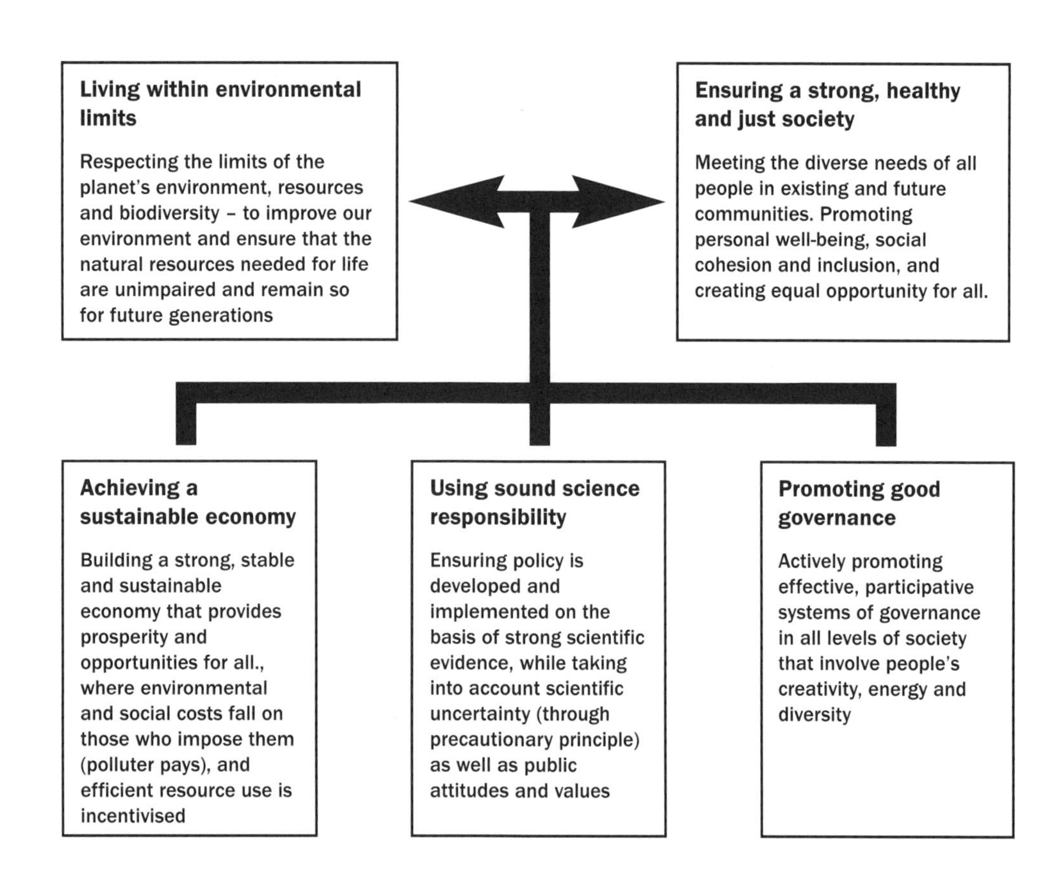

Figure 3.1 *Shared UK principles of sustainable development (source Defra, 2008b)*

3.4 HOW IS SUSTAINABILITY LINKED TO WATER CYCLE MANAGEMENT?

The sustainable consumption of water, disposal of wastewater and management of surface water is important for the long-term health of the environment, economy and society. *Future water* (Defra, 2008a) has highlighted ways to achieve sustainable water use by reducing and limiting consumption. This includes a reduction in the environmental effects of abstracting, distributing and treating the water we drink and collecting and treating our wastewater before it is returned to the natural environment.

The demand for water is increasing, resulting in high abstraction levels in some areas of the UK. This puts significant strain on the environment, especially during drier periods. Falling groundwater and river levels, and the drying of wetlands can lead to irreversible biodiversity loss and even threaten non-freshwater habitats, for example, where tree roots tap into the water table. The treatment of wastewater and effects of combined sewer overflows during and after storm events also has a significant effect on our environment, so their management is vital in achieving SWCM.

The reduction of carbon is at the forefront for achieving sustainability. The links between carbon and water are becoming more evident. So there is a strong link between water cycle management and sustainability. This is evident as the current water cycle management uses large amounts of carbon as discussed in Chapter 2. The high energy consumption of the water industry highlights that a more sustainable approach to water cycle management will reduce water waste and carbon emissions.

Overall, sustainable water cycle management involves making decisions to meet the current needs of society, industry and the environment in a resilient, adaptive and productive manner rather than in a vulnerable and destructive fashion.

3.5 HOW CAN SUSTAINABLE WATER CYCLE MANAGEMENT BE ACHIEVED?

3.5.1 Unsustainable development

Understanding unsustainable development permits us to have an improved appreciation of how sustainable development can be achieved. Waterwise states that the current average per capita rate of consumption of 150 l/p/d in the UK is unsustainable in the long-term because of increasing water stress (Waterwise, 2009). Climate change, urban population growth and changes in behaviour all mean that the UK is increasingly at risk of water stress in future.

In the past, little consideration was given to managing flood risk, surface water management and water efficiency. This led to unsustainable methods of growth where:

- developments have been sited in high flood risk areas
- new developments potentially affect downstream flood risk
- for surface water runoff greater demand has been placed on existing infrastructure and in some cases, capacity is regularly exceeded
- infrastructure and conventional treatment methods often require a significant amount of energy for pumping
- habitats have been adversely affected by new developments
- there has been little concerted effort to reduce demand for water.

Water has traditionally been regarded as a plentiful resource, and the locating and design of new developments has given little weighting to the local availability of water. Under the Water Industry Act 1991 as principally amended by the Water Industry Act 1999 and the Water Act 2003, water utilities have had a duty to supply water to new developments that then become their customers and add to their revenues.

The approach of transporting water off-site through the quickest means possible exacerbated pollution incidents and potentially contributed to higher flood risk downstream. This combination of increasing demand and high runoff rates caused rising volumes of water entering the system leading to overloading of infrastructure systems and treatment plants. Managing flood risk, surface water management and water efficiency should be integrated during the planning and delivery of sustainable developments.

3.5.2 Sustainable development

Sustainability is difficult to define as an outcome and due to our consumptive nature as humans, full sustainability may be impossible to achieve. However, when sustainability is considered as a process of learning and communicating, ie a journey rather than a specific end goal, it becomes much more achievable. Sustainable development can be defined as a combination of activities and processes. Four key areas have been identified by the UK Government:

1. Sustainable consumption and production – using products and services efficiently by achieving more with less.
2. Climate change and energy – reducing greenhouse gas consumption.
3. Natural resources – using resources such as water, air and soil wisely.
4. Sustainable communities – looking after the places people live and work.

There are no generic criteria applicable to each and every development to ensure sustainability, but there are generic ways of working that encourage sustainability. One of these is early consultation with the stakeholders affected by the development, to assist with informed decision making. Sustainable development as a process can be achieved through the relevant stakeholders working together to identify the opportunities and constraints, both at a local and regional level, and working within these constraints to develop solutions that realise the multiple benefits from the opportunities. For example, if a large body of water needs to be stored on-site, this may not need to be held in a large underground storage tank but rather it could be considered as an above-ground pond that would create biodiversity and amenity benefits.

Different methods and technologies to help achieve sustainable development should be adopted depending on several factors including the scale of the site, the local situation and the regional water issues. The examples below highlight the varying scales at which these systems can be delivered and examines the different methods of water management integrated into the developments depending on their regional location. The Millennium Green Project in Nottinghamshire concentrates on water efficiency, the Dunfermline Eastern Expansion in Scotland focuses on pollution control and reducing flood risk and the Elvetham Heath project uses a combination of methods to manage surface water runoff and save water.

Case study 3.1 *Millennium Green, Nottinghamshire*

Millennium Green is a small scheme of just 40 residential units and an office. The site uses an automated rainwater system that is topped up with mains water if necessary. The system supplies water for toilet flushing, washing machines and watering gardens. The development was awarded the Environment Agency's 2003 water efficiency prize and has reduced the sites water use by 50 per cent a year.

Case study 3.2 *Elvetham Heath, Hampshire*

Elvetham Heath is a 126 ha development situated in the south-east of England. It contains a natural reserve, 62 ha site planned to accommodate 6000 residents and car parks. A combination of SUDS and water-saving measures were used to improve water sustainability on site.

SUDS were incorporated for the village runoff and included swales/retention ponds that help manage flood risk and also contribute to the aesthetic value of the village. Dual-flushing toilets, low-flow showerheads and efficient washing machines helped reduce water consumption and improve efficiency. The urban water orienteering tool (UWOT) (Chapter 6) identified these as being the best water management practices for Elvetham Heath.

Case study 3.3 *Dunfermline Eastern Expansion, Scotland*

> The Dunfermline Eastern Expansion (DEX) is a 550 ha (5.5 km^2) site that lies to the east of Dunfermline. The site, which was predominantly greenfield, will be developed over the next 20 years as a mixture of industrial, commercial, residential and recreational areas.
>
> The main concern for this site was the quality and volume of runoff to the river further down the watercourse. These issues were further exacerbated because the downstream river had an existing flood risk problem. To mitigate these effects SUDS were adopted and regional SUDS infrastructure was put in place before development started.
>
> DEX is located on an area of predominately low-permeability clay soil. Infiltration methods have limited application on the site. Some residential roads will be served by soakaways, where soil permeability permits. Much of the spine road system is drained using offlet kerbs, filter drains and swales, which discharge into extended detention basins and wetlands that also serve adjoining housing areas. Treatment of surface water runoff from the development and roads is achieved through a system of regional ponds and wetlands before discharge to the watercourses. Ponds and basins are widely used to achieve maximum attenuation of storm flows. The wetland is located in a public park area where informal public open space adjoins an existing forested area and an area set aside for football pitches, a rugby pitch and tennis courts. Permeable paving has been used in a supermarket car park, which is connected through attenuation/infiltration basins to the wetland.

These case studies demonstrate how different outputs can improve the sustainability of water cycle management. They highlight how sustainable development is the process of identifying the relevant site-specific issues and then forming site-specific water management objectives.

Water and sustainable development

Water can help create sustainable places. It is a valuable asset that can bring economic and health benefits, which improve people's quality of life. Sustainable water management can provide social benefits for a development including supporting attractive and valuable habitats that can improve biodiversity.

Residents can benefit from water being incorporated into the landscapes. It is generally considered that the health and well-being of the residents is improved as these environments create opportunities for sport, recreation and relaxation.

3.5.3 Sustainability and water planning

The planning system is instrumental in achieving sustainable developments. Many Planning Policy Statements emphasise the importance of environmental sustainability and of delivering suitable water infrastructure to achieve it. Sustainability objectives need to be considered at all stages of the planning process.

Within the spatial planning system the core strategy sustainability assessment, the sustainability appraisal of the regional plan and water cycle study guidance (WCS) provide the necessary basis to generate sustainability objectives at a strategic level. The latter stage of WCS, the Code for Sustainable Homes (water-efficiency targets) and the planning permissions process focuses on accomplishing SWCM in new developments.

Water cycle studies

Water cycle studies are a relatively new approach that supports sustainability in the urban environment with respect to managing the urban water cycle. WCS provide a more sustainable solution to the issues of growth, climate change and tightening standards (WFD).

The purpose of the WCS is to examine the potential effects of future growth in relation to three main aspects of the water cycle: water resources, water quality and flood risk. The strategy will ensure planned development is within environmental capacity by:

- identifying where and when water cycle related infrastructure is needed to support housing delivery
- agreeing development policies to be applied by the planning authority to achieve the most sustainable solutions
- ensuring ongoing management and maintenance of infrastructure.

To coincide with the planning process a WCS is broken down into three stages; the scoping study, the outline study and the detailed water cycle. See Appendix A3.1 for further information.

The WCS help planning authorities make strategic decisions about land allocation for development and what management options should be included. This guidance sets out the tools to help make decisions on a site-specific basis about the sustainability of solutions and provides the evidence base for a planning authority to support them through the WCS.

Code for Sustainable Homes (CSH)

The CSH sets out water efficiency targets for new homes by outlining different levels of water consumption. The Code is voluntary for privately built homes but for all new social housing Code level 3 should be achieved. Table 3.1 highlights the methods necessary to meet each level.

Table 3.1 *Achieving the Code for Sustainable Homes*

Water consumption (litres/person/day)	Credits	Level	How this is achieved
≤120	1	1 and 2	Basic water efficiency including low-flush toilets and aerated shower heads
≤110	2		
≤105	3	3 and 4	Water efficiency and the use of RWH systems
≤90	4		
≤80	5	5 and 6	A more comprehensive rainwater system is necessary or a greywater system

The CSH and the latter stages of the WSC provide indicators for achieving sustainable development on a site scale. At a policy level, the benefits of sustainable development may be irrefutable however the translation into practice is not straightforward. The development of criteria and monitoring tools are important to deliver changes and help wider societal change. Adaptation to changing circumstances is fundamental to a more sustainable approach, both in the way decisions are made and how solutions are delivered. The integration between different disciplines as well as those affected by the decision is a main principle of sustainable development.

3.5.4 Sustainability: principles, criteria, indicators

Sustainable principles, criteria and indicators are used at the master planning and delivery stage for water cycle management. They act as a checklist for the delivery of SWCM, establish the different priorities of stakeholders and help the decision making process. They should be considered from the preliminary stage and consulted at all stages of the development process.

Several example principles, criteria and indicators are available. The Sustainable Water industry Asset Resource Decisions (SWARD) (Ashley *et al*, 2004) and the Sustainable Project Appraisal Routine (SPeAR®, developed by Arup) were exploited in the WaND project for their wider issues of sustainability.

SWARD was a collaborative project between UK academics and water service providers in Scotland, England and Romania. It was developed as a practical tool to assist with the explicit inclusion of sustainability in the decision making process for those responsible for providing water services. SPeAR® was developed to demonstrate the sustainability of a project, process or product. A framework of criteria and indicators for achieving sustainability was developed to bring the issues of sustainability into the decision making process. In these frameworks the principles of sustainability are simplified into constituent categories of criteria that need to be equitably fulfilled to provide a balanced decision. A useful checklist is also included in guidance published by the Environment Agency (2006).

The SWARD procedure focuses on the incorporation of sustainability criteria into investment decisions and by presenting an inclusive and generic set of sustainability criteria for use in the water industry's decision making processes. The SWARD procedure distinguishes between principles, criteria and indicators:

- **sustainability principles** are definitions, or goals for sustainability, which should remain constant over time and may be the same for different cases (Upham 2000). For example "better quality of life for everyone, now and for generations to come" (Woods-Ballard *et al*, 1987)
- **sustainability criteria** are factors that may be used to assess the range of options that can offer the greatest contribution to achieving sustainability objectives, as described by the principles (Foxon *et al*, 2002). The criteria generally falls into, social, environmental and technical criteria
- **sustainability indicators** are variables, used to measure and assess the sustainability of a system or for comparing between different solutions (Balkema, 2003) and are associated with a specific sustainability criterion as a means for their assessment. The criteria are subdivided to include measurable indicators such as cost and labour, extraction and emissions, durability and maintenance and participation and local development. Further information and a full list of indicators can be found in the literature (see Azar *et al*, 1996, Butler and Parkinson, 1997 and Butler *et al*, 2005).

For each new development, it is likely that new indicators and possibly new criteria have to be developed. The following table provides an example of criteria for a site. WaND used the SWARD decision making framework that included four headline sustainability categories (economic, social, environmental and technical) plus a series of associated, generic, primary criteria (Table 3.2) as well as user-defined, specific, secondary criteria and indicators.

Table 3.2 *Categories and primary criteria of sustainability evaluation used by SWARD*

Category	Primary criteria
Economic	• life cycle costs, the whole-life costs of investment, resource extraction, production, construction, end use and decommissioning • willingness to pay for a stated attribute incorporated at design stage • affordability to customer • financial risk exposure, the risk of loss associated with investment.
Environmental	• resource use, the land, energy, chemical, material and water resource use during the whole life of investment • service provision: water consumption, leakage and reuse • environmental impact on land, water, air and biodiversity.
Social	• effect or risks to human health • acceptability to stakeholders of the investment scheme • participation and responsibility: stakeholders participation and responsibility in sustainable behaviour • public awareness and understanding: awareness of sustainable development and implications of behaviour • social inclusion by water utility actions.
Technical	• performance of the system, quality of effluent treated in relation to required standards • reliability of the system in performing its functions • durability: the level of accommodation in design of the system • flexibility and adaptability: the ability to add or remove from the system.

The SPeAR® framework, which was also used in WaND, is based on a four quadrant model that structures the issues of sustainability and allows an appraisal of performance to be undertaken. It focuses on the main elements of environmental protection, social equity, economic viability and efficient use of natural resources. Table 3.3 outlines the categories and criteria used in the framework where sustainability is achieved by each category being equally weighted.

Table 3.3 *Categories and primary criteria for a sustainable appraisal used by SPeAR®*

Category	Primary criteria
Environmental	♦ air quality ♦ land-use ♦ water discharge ♦ natural and cultural heritage ♦ design and operation ♦ transport.
Social	♦ health and well-being ♦ stakeholder engagement ♦ form and space ♦ access ♦ amenity ♦ social responsibility.
Economic	♦ transport ♦ employment skills ♦ competition effects ♦ viability.
Natural resources	♦ water hierarchy ♦ energy ♦ water use ♦ materials.

The Environment Agency's guide for developers (EA, 2006) provides a more comprehensive set of principles. This includes 10 principles and several actions for each principle. Three focus on SWCM:

- managing the risk of flooding
- managing surface water
- using water wisely.

They are aimed at developers but provide a good indication of factors that should be considered in a new development.

3.6 WATER MANAGEMENT DECISIONS FOR THE FUTURE

Scenario planning and descriptions of alternative theoretical futures that reflect different perspectives on developments can serve as a basis for planning and action.

Scenarios do not predict the future but are used as tools for thinking about it and to assess the effects of a range of possible decisions. They provide representations of multiple possible (and not necessarily probable) futures through a storyline (or several storylines) that are plausible.

Futures scenarios help in the collective processes of deliberation involving professionals and other stakeholders. Their value lies as problem solving and decision making tools for establishing thoughts on the future and sustainability.

Scenarios were used within the WaND research project to explore the longevity and adaptability of solutions compared to the conventional alternative. Scenarios should become an integral part of decision making for urban water management. They allow a greater understanding of what the future pressures could be on the water cycle to adapt, plan and provide for these changes. For a further discussion on developing scenarios see Appendix A3.2.

3.6.1 Using the scenarios

The authors of *Foresight future flooding* suggest two approaches (Evans *et al*, 2004):

1 Use the scenarios to stimulate thought on what the future holds and to consider the implications for medium and long-term strategies.

2 Use the scenarios as the basis for a sensitivity-analysis type study on a specific sector or issue.

In this approach, the scenarios provide the system boundaries within which the sector's long-term strategies are assessed. The main challenge is combining the soft scenario tool with hard quantitative methods.

The scenario framework is a flexible tool that can be adapted and altered to suit the needs of any study. Identifying the main objectives in the sector is required and, in the case of water demand management (EA, 2001b), scenarios were developed for specific micro-components of public water supply (eg household, leakage, non-household). Table 3.4 provides examples of the common drivers that can be used at all stages of SWCM. Classification has been identified from the sustainable categories used in SWARD. The drivers and description explain the development scope towards or away from sustainability.

These scenarios can be used in two distinct ways within the context of SWCM:

1 As a platform for exploring the effect of alternative socio-economic, technological or environmental situations on suggested solutions (technological, institutional or legislative options produced through guidelines or tools), particularly if the solutions are strategic in nature and are required to endure for timescales relevant to the scenarios (ie 10 to 50 years).

2 To develop different user preferences for use with screening and technology optioneering tools (Makropoulos *et al*, 2008b), investigating the effect of perceptions under different socio-economic situations to, for example, the siting of new urban developments and selecting appropriate technology sets to support (robust) SWCM in these developments.

Table 3.4 *Objectives for SWCM*

Classification	Objective	Description
Cross-cutting	Regulations	National or international legislation and regulations affecting SWCM
Societal	Demography – population growth	Demographic characteristics and temporal trends affecting SWCM
	Settlement patterns	Characteristics of urban development including spatial trends and building concepts and strategies
	Domestic habits	Habits of the water users within their private/home environment
	Lifestyle, values and perceptions	Values and beliefs governing everyday life and urban water in particular
Economic	Cost of service including cost of resource	Cost for the provision of the SWCM service taking into account the cost of water or resources needed for its management as a measure of its financial importance
	Cost of failure	Cost from failure of the system (ie failure of provision of the SWCM service)
	Disposable income	Income available to the end users to "do as they like" after having satisfied their immediate needs
	Investment priorities	Priorities for investment for companies and customers related to SWCM
	Ownership issues	Ownership of service and/or assets.
Environmental	Resource availability and climate change	Availability of the individual water flow, seen as a resource, taking into account the potential long-term, significant change in climate and its effect on SWCM
	Sustainability agenda	The attitude (public and private) towards the concepts of sustainability and sustainable development

An example of a scenario approach to urban water planning, is included in the EA's work on options for distributed water infrastructure (Butler and Makropoulos, 2006), where a series of water management systems was modelled to investigate the effect of centralised and distributed infrastructure options to urban water management. The potential effect of the identified infrastructure options was examined through the selection and assessment of indicative scenarios. These scenarios (Table 3.5) assessed different sets of compatible technologies and illustrated their effect on the urban water cycle under different situations.

The benchmark scenario depicts the current state of legislation and regulation. The next-step scenario is the most realistic for the short-term. The last two scenarios are more radical and attempt to identify technologically feasible options for addressing water infrastructure constraints that may appear in the context of specific future urban development projects.

Table 3.5 *Distributed infrastructure scenarios*

Realistic scenarios	
Scenario 1: benchmark scenario	Practice under current Building Regulations
Scenario 2: next step scenario	Delivery of a full set of water-efficiency measures, metering and source control and use of non-potable water from rainwater or local groundwater
Radical scenarios	
Scenario 3: limited central infrastructure scenario	Introduction of greywater recycling and small-bore sewers
Scenario 4: no central infrastructure scenario	Emphasis on local solutions and water autonomy

The desirability of these scenarios is subject to constraints, including water resource availability, treatment plant capacity, cost of upgrading infrastructure, developments in (distributed) energy (micro) generation and climate change. Assessment of the scenarios is only indicative, as the performance of the technological options is associated, to a large extent, with case-specific characteristics and cannot be easily generalised.

The results for the three scenarios, in comparison with Scenario 1 (benchmark), are shown in Figure 3.2. All the scenarios improve across a series of indicators compared to the benchmark, such as savings in potable water consumption. Even more drastic improvements are proven to be possible in reductions in (external) water supply and reductions in rainfall runoff leaving the site.

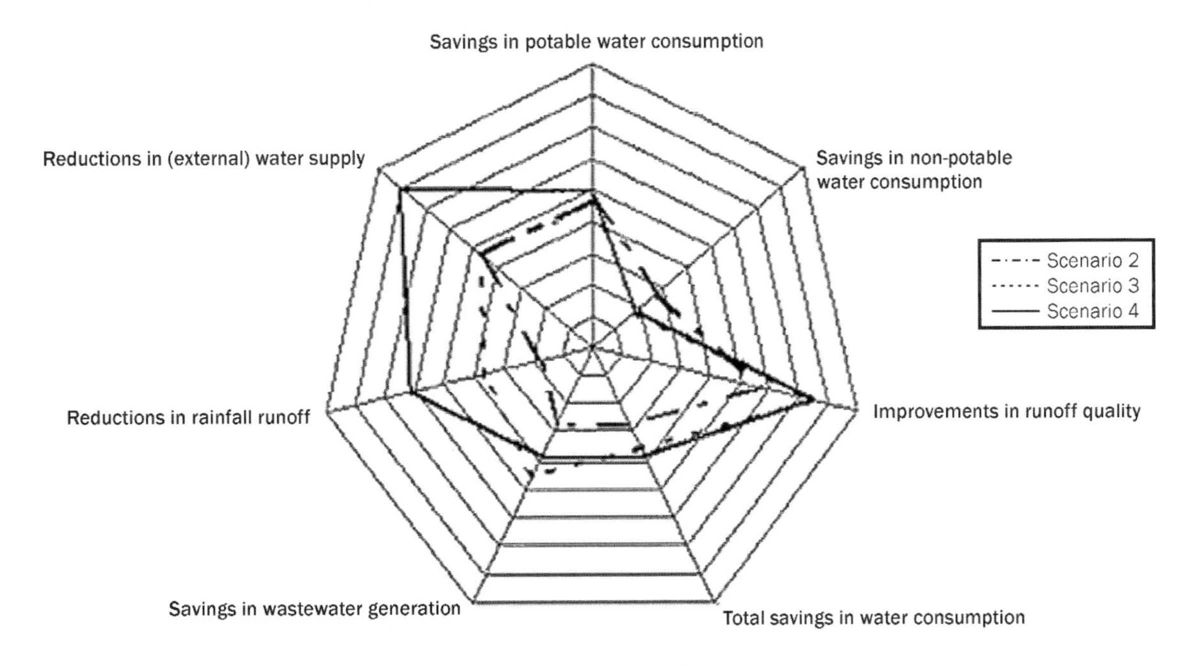

Figure 3.2 *Scenario comparisons in terms of water (as per cent improvement from benchmark)*

The scenario exploration suggested that there is considerable room for improvement in urban water management, even when deploying fairly standard, understood technologies, across a series of criteria. However, it showed that even for more extreme situations, technological options can be deployed. It may also be suggested that scenarios such as the ones developed within WaND, could act as a mechanism to foster further technological innovation in particular fields, but develop a common understanding of future possibilities and potential challenges.

3.7 CONCLUSIONS

- it is evident that our current trend of water management practices is unsustainable
- sustainable water management is critical to ensure future water demands are met, especially with the predicted increase in demand and the effects of climate change
- in many cases we have reached or are nearing the capacity of our water infrastructure so water efficiency, localised water treatment and surface water management are fundamental in the planning process
- sustainable development requires a more inclusive approach to decision making to ensure new developments have few environmental effects, improve the environment where possible and create sustainable places environmentally, economically and socially. If water is managed creatively it can have multiple benefits
- sustainability principles, criteria and indicators aid the decision making process and allow an inclusive approach for all stakeholders
- water cycle studies (WSC) help achieve the most sustainable solutions for water management through planning policies and ensuring the right infrastructure is delivered when necessary in the development process
- SWCM needs to focus on regional-scale constraints and objectives to achieve the most effective outcomes, as priorities across regions will vary
- more sustainable water cycle management can only come about by the integration of all those involved as current practices are mostly too insular and institutionalised
- water cycle management in a new development is linked to management of the whole development and its place in the catchment. New ways of working are required that allocate time and money for public engagement and cross-disciplinary consultation.

3.8 FURTHER RESEARCH AND GUIDANCE

- **this chapter sets out the different aspects that need to be considered but does not give a clear set of principles, indicators or criteria**
- **case studies identify examples of good practice but SWCM is site-specific**
- **a better understanding of the long-term issues associated with costs and maintenance of certain types of sustainable water products.**

4 Technologies

This chapter:

- outlines some of the technologies used in water supply management, stormwater management, wastewater management and water recycling/reuse
- sets out some of the technologies for water demand management and stormwater management, showing potential to help deliver sustainable water cycle management (SWCM) in new developments
- summarises some of the available water-saving devices for household usage including low-flush toilets, aerated showers, efficient washing machines etc
- presents an overview of rainwater harvesting (RWH) systems to help water supply management and stormwater management
- highlights some of the different treatment processes associated with greywater recycling (GWR) and presents findings of research into five of the treatment processes available
- illustrates the benefits of SUDS for stormwater management.

4.1 INTRODUCTION

There is an extensive range of technologies available to help SWCM. This guidance summarises some of these technologies and identifies the most sustainable scale for their application. It seeks to highlight how water can be managed sustainably at different stages of the water cycle using various technologies.

This chapter is not a comprehensive guide on which technologies to choose, or how to implement them, but rather provides a summary of what was reviewed during the WaND research project, what is available on the market and what is emerging in the research and development arena. This should enable stakeholders to make informed decisions on appropriateness of technologies, the scale they should be used at and provide an indication of the cost they may entail.

The technologies identified are organised into four groups: water supply management, rainwater harvesting (RWH), greywater recycling (GWR) and stormwater management (Tables A4.1a to A4.1d in Appendix A4.1). These relate to the three urban water flows (water supply, stormwater and wastewater), although the focus of the chapter is on water supply and stormwater. There is some degree of overlap between the technologies, for example, RWH reduces demand and can help stormwater management by alleviating large runoff rates.

This guidance provides a slightly more detailed look into the technologies researched within the WaND project. The WaND research project mainly focused on:

- investigating and forecasting the influence of water-saving technologies on overall water demand at a range of scales (local, regional and national)
- investigating the performance and influence of alternative water supply systems (eg greywater recycling)
- investigating the performance of innovative wastewater collection systems and developing design/installation guidelines.

The WaND research project identified a wide range of traditional and novel technologies that have been used in different parts of the UK, as well as overseas. These are briefly described in Appendix A4. Tables A4.1a to A4.1d and Box A4.1 then presents a selection of guidance documents that cover different aspects of SWCM in relation to technologies. A summary of the main outputs and findings is presented in the following sections.

4.2 WATER DEMAND MANAGEMENT

A significant proportion of new homes are expected to be built in south-east England, which is already regarded as water stressed (Figure 2.1). *Future water* (Defra, 2008a) continues to support a twin-track approach to ensure future security of supply. This includes the use of demand management measures targeted at conserving water and reducing per capita consumption with rigid water resource planning to meet future demand.

To encourage efficient use of resources in England and Northern Ireland, the Government has introduced the Code for Sustainable Homes (CSH). The CSH water efficiency related targets are summarised in Table 3.1. All new social housing needs to be at level 3 or above. The Code is voluntary for privately built housing, but houses built to the Code should have a mandatory rating. Detailed guidance on how to achieve these targets is given in the accompanying technical guide (CLG, 2008).

The reduction in per capita consumption needs to be achieved through the delivery of water demand management measures. These include water saving devices, greywater recycling and rainwater harvesting.

4.2.1 Water-saving devices

The average per capita daily water consumption in households is about 150 litres. Table A4.2 shows the current volumes of water used in household appliances.

These volumes can be reduced considerably by installing low-water using appliances. The best available technologies not entailing excessive costs (BATNEEC) lists how consumption can be reduced to 76 litres per person per day using 2001 appliances and 53 l/p/d using 2006 appliances (see Table A4.3).

The potential savings (Table A4.3), for each appliance are theoretical in nature and require validation. To address this, several water companies have conducted a range of water-efficiency trials to gather evidence. The preliminary findings can be found in *Sustainability of water efficiency* (UKWIR, 2006a).

The effectiveness of a range of water demand management options (eg metering, low-flush toilets, rainwater harvesting) was analysed in the WaND research project. The analysis was intended to give an indication as to which conservation options have the greatest effect in moderating domestic water demand across the UK. Though domestic consumption is likely to rise in the medium term, conservation options still play an important role in reducing the rate of increase.

A summary of comparative analysis is shown in Table A4.4. The table relates to the potential reduction of water demand by several water-saving technologies compared with standard UK appliances. The table shows the potential amount of water each technology can save and estimates the likely national uptake of the device. For a water device to significantly reduce national water demand it needs to save water and be adopted by a significant amount of UK households.

Low-flush toilets, normal-flow showers and reduced-flow basin taps per capita consumption can conservatively be reduced (by 30 lpd). The reduction potential could be much higher if other water-saving devices are used on appropriate scales such as greywater systems at a community level. This is particularly true for new developments.

The WaND research also investigated the link between water saving devices, CO_2 emissions and sustainability, highlighting the water saving and CO_2 emissions associated with some shower types in Table A4.6. Also for different types of taps and WCs flush volumes the sustainability advantages and disadvantages were explored in Tables A4.6 and A4.7.

Further details on all the water management options discussed including information on their operating mechanisms, anticipated water efficiency and cost-benefit analysis can be found in the Environments Agency's review on water conservation measures (EA, 1999). Up-to-date guidance on water saving technologies and approaches can be found in the practical guide from the Environments Agency (2007b), which includes information on water-efficient:

- WCs
- urinals
- showers and baths
- white goods (washing machines and dishwashers)
- taps
- gardening
- reusing greywater and harvesting rain
- detecting leaks and metering
- plumbing and heating systems design.

The WaND research specifically focused on the potential of low-flush toilets as a water saving device.

Low-flush toilets

About one third of household water use is for flushing the toilet (with typical flush volume of nine litres). An effective flush volume is the volume of water needed to clear the toilet pan and transport solids far enough to avoid blocking the drain. Too little water will lead to double flushing and increased risk of the drain blocking, while too much will waste water (EA, 2007b). Water Fittings Regulations (1999) made it mandatory that all new toilets need to have a flush volume not exceeding six litres. Also 6/4 litre dual-flush toilets are now allowed.

To address the design of toilets with low-flush volumes, the WaND research project developed and investigated a small-bore wastewater collection system (see Box A4.2).

The research highlighted that in the future reducing water consumption through installation of low-flush WCs is envisaged to be one of the most cost-effective interventions. The wastewater collection designed in WaND is capable of effectively flushing the WC waste and with no blockages so far. The likely wider uptake of the system is dependent on modifications in the Building Regulations to accommodate connectivity to small diameter pipes.

Costs

The selection of water-use fixtures and fittings is usually a subjective (design) matter linked more with affordability than actual water use. Low-water use fittings and appliances are generally very similar to conventional products and are likely to display similar price variations and ranges when they come into mass production, meaning extra costs are unlikely. Further potential issues related to costs (eg plumbing) are outlined in the introductory guidance by Griggs and Burns (2008).

Future

The longer-term targets of the CSH can be met and it is possible to considerably reduce per capita consumption using water-saving devices. The Environment Agency guide on harvesting rainwater for domestic uses (2008a) also supports installation of water-saving components as a preferred option. The likely uptake of water-saving devices depends on the effectiveness of policy and changes in the socio-economic make-up of the country over time.

The WaND research project developed several demand forecasting tools to capture the influence of most of the major factors on the overall demand including:

- population growth
- household size
- lifestyle
- climate change
- use of new water-saving technologies.

The major factors were looked at on different scales (household, ward, local authority regional level) and time horizons. These can inform the decision-support process for water planning in future developments. These tools are discussed in Chapter 6.

4.2.2 Rainwater harvesting (RWH)

Rainwater collected (or harvested) from roofs and other hard-standing areas (driveways), after some basic treatment (eg filtration/storage), can be used for non-potable applications, especially toilet flushing and garden irrigation. The use of RWH systems to supplement potable supply is widespread (see Box 4.1). In the UK, the market for RWH systems is growing with nearly 1000 domestic and 300 commercial systems sold in 2007 alone. According to the UK Rainwater Harvesting Association (UK-RHA), which accounts for about three quarters of the market, the overall size is about £5m in comparison with a German market worth around £1bn a year (EA, 2007c).

Box 4.1 *RWH as water demand management (WDM) option – scale of use*

- in Australia, more than one million people rely on rainwater for potable use
- in the USA, 20 000 systems have been installed to meet domestic water needs in rural areas
- in Germany, over 100 000 systems have been installed in low and high density buildings to meet non-potable demand.

In new homes, around 35 per cent of the total water used can potentially be substituted by rainwater (water used for toilet flushing and some external uses). In a house with four people, this amounts to a potential saving of about 207 litres per day, or 75 m^3 a year. However, for a system to deliver these savings, it would require a large collection area and high rainfall (EA, 2008a).

Factors influencing the effectiveness of RWH systems include:

- rainfall patterns
- catchment (roof, hard-standing) area
- storage volume and duration
- water demand patterns.

These factors have a direct impact on choice of tank size. Several approaches, eg Fewkes (2006) and Environment Agency (2008a), are available to determine optimal tank size and a detailed guidance describing the RWH system design procedure, installation considerations and cost-benefit analysis has been prepared by the Environment Agency (2008a).

RWH systems can be used in a range of settings including residential blocks, commercial buildings (office blocks, hotels, schools, hospitals, factories, warehouses and retail sheds) and even a small number of community schemes. Box A4.3 provides two examples of RWH systems used in the UK at two different scales. BS8515:2009 on rainwater harvesting also provides detailed guidance.

Costs

The overall cost of an installed RWH system can be divided into several components in addition to the capital cost of the basic system (tanks, pipework, pumps, filters, disinfection and controls):

- preparatory works (excavations/plinths for tanks etc)
- collection pipework and components
- distribution pipework and components
- installation and commissioning
- redecoration (if retrofitting)
- maintenance (filters and disinfectants (eg UV lamp replacement) and power charges). Important maintenance activities are listed in Table A4.9.

Clearly, the installation costs would be much less for a new building where RWH can be incorporated into the design. In retrofit applications, other than where the existing layout is particularly favourable, installation costs might well exceed the hardware costs of the basic system. Detailed guidance on factors influencing the system costs and costing methodology is given in Leggett *et al* (2001a). Further details of cost benefits analysis on the systems used in the WaND research can be found in Box A4.3. Also payback periods of RWH schemes are highlighted in Table A4.11.

The cost of installation and maintenance of RWH systems is a key feature when deciding if a system should be used. A detailed review of the technical and economic aspects of RWH systems at various scales and in diverse settings is provided by the Market Transformation Programme on rainwater and greywater (MTP, 2007). The MTP document and Environment Agency guidance suggests consideration of the following points when delivering RWH systems:

- in affordable low-rise housing, the rainwater reuse is constrained by roof area where the minimum living space per person (based on current Section 106 planning guidance) is 30 m^2, providing only 15 m^2 of roof area per person
- using average rainfall in England and Wales (905 mm/yr) and assuming a system

efficiency of 60 per cent, a roof area of 138 m^2 would be required to meet an annual demand of 75 m^3. In areas that experience less than average rainfall this roof area will need to be bigger to satisfy the demand for rainwater. This suggests that the benefits of rainwater harvesting will be small in modest homes in areas with low rainfall such as eastern England and significant in larger homes in areas with abundant rainfall such as north-west England

- only larger or less densely occupied private housing is likely to achieve considerable (about 50 per cent) mains-water savings, unless surface-level catchment is also considered
- where there is insufficient rainwater to supply all non-potable uses, distribution can be limited to one outlet, which minimises pipework costs
- the roof area of a typical single household is not sufficient to meet non-potable water needs fully. A two or four house system is only slightly more expensive than a single-house system. One alternative is to collect rainwater from two houses but only supply it to one
- factories, warehouses and retail sheds usually have sufficient roof area to easily meet WC-flushing requirements for staff and/or customers
- factories should also consider whether rainwater can be used for process needs. A good application for rainwater in factories is make-up water for cooling towers and steam boilers where the very low level of hardness and dissolved solids reduces water treatment costs
- in situations where large and lightly populated buildings can only use a small proportion of the rainwater resource, consideration should be given to sharing the rainwater resource through a community scheme (MTP, 2007).

Future

Using rainwater for non-potable applications requires minimal treatment and acts as a water demand management option. It can reduce per capita potable water consumption by about 35 per cent. Looking at UK-RHA business trends, it is envisaged that uptake is likely to grow significantly in the future. The main considerations for the using RWH systems are location to ensure rainfall yields, and the building type to ensure optimum use of the system.

4.2.3 Greywater reuse

Greywater is wastewater collected from baths, showers and washbasins (and in some cases washing machines). It excludes kitchen and WC waste due to their high organic content, which is known as black water. Greywater is less polluted than black water and its quality depends on several factors including, the lifestyle of the producer (personal washing habits and products used) and climate. Once the greywater is collected it can be treated and reused for toilet flushing, heating systems, garden irrigation, car washing, possibly washing machines (if the water is treated to a high standard) and any other use that potable water quality is not required for. There is still some debate on how greywater can be used for irrigation, with many practitioners preferring to use drip-feed irrigation systems on non-edible plants.

Greywater treatment systems

Greywater systems (GWS) can vary significantly in their complexity and size. The common features in most greywater systems are:

- a form of treatment facility (filtration and disinfection)
- a tank for storing the treated water
- a pump
- a distribution system to transport the water to where it is needed.

All GWS that store water have to incorporate some level of treatment, as untreated greywater deteriorates quickly. The extent of treatment depends on several factors including the quality of incoming greywater and the scale of reuse.

There are several GWS that can be categorised according to the type of treatment they use. The different treatments include:

- **direct-reuse system (no treatment):** if greywater storage is short then no treatment is necessary. Very simple devices can make this practical, for example, bath water once cooled can be used for watering the garden
- **short-retention system:** a very basic treatment is used, the greywater settles and the debris is skimmed out. The main feature is the short storage time of the water and its simplicity reduces the need for maintenance. The water is stored for no longer than 24 hours and after this time period any water is removed and treated as waste
- **chemical systems:** a filter removes the debris, the greywater is then stored and a chemical is applied to stop bacteria growing. This system allows for longer storage periods of the greywater but has higher cost and requires more maintenance
- **biological systems:** bacteria are used to remove organic matter by the introduction of oxygen. The systems vary in complexity and form, which has implications on their costs and maintenance
- **bio-mechanical systems:** a combination of biological and physical treatments are used to provide a higher standard of water, which complies with EU bathing water standards. The greywater undergoes an extensive cleaning regime. Organic matter is removed, solid material settles before being removed automatically, oxygen bubbles through the water to encourage bacterial activity and finally it undergoes UV disinfection.

The variations in treatment depend on the scale of their application and required end use. For single households, where exposure to greywater is likely to be minimal, a simple form of treatment usually a chemical system is normally required before greywater is used for toilet flushing. A variety of small-scale, portable, packaged systems with control panels to monitor system performance and help maintenance are available in the market. For large-scale systems (eg a block of flats), further treatment (mainly biological) is commonly practiced. These systems are employed to minimise health risks, in countries such as Australia, Japan and Singapore.

LANDCOM, an Australian based organisation, has compiled a list of treatment technologies and their associated attributes including, capital and operating cost, effluent water quality and suggested uses, space requirements and recommended development scales (single household, small, medium or large development). A full list is provided in Appendix A6.

The main treatment employed to remove organic pollutants is a biological treatment process, which can be sensitive to the nature and concentration of the pollutants discharged. The WaND research project undertook a survey to investigate the frequency of different substances discharged into greywater and the substances that are perceived as being harmful to the environment. The findings from the study on substances discharged into greywater can be found in Appendix A4.6.

WaND research – greywater treatment technologies

The WaND research project investigated the suitability of five technologies for the treatment of greywater. These were:

- biological system: a membrane bioreactor (MBR)
- chemical system: a membrane chemical reactor (MCR) based on an advanced oxidation process (TiO_2/UV)
- horizontal-flow reed bed (HFRB)
- vertical-flow reed bed (VFRB)
- "GROW" green roof water recycling system a novel system.

These technologies were reviewed as they were considered innovative at the time and were thought to contribute to potential approaches for the future. A brief description of these technologies with respect to their construction and operating conditions is given in Boxes A4.4 to A4.7. Further information can be found in Pidou *et al* (2007).

In this study, the GROW and the HFRB systems achieved limited treatment of the greywater. The MBR, the MCR and the VFRB systems achieved good general treatment of the greywater. However, the MBR and MCR alter the solids and microbial fractions more significantly.

Overall, the MBR was found to be the most suitable technology for greywater recycling due to its overall robustness. Indeed, the MBR constantly achieved excellent treatment on high concentrations of greywater. MBR is an energy-intensive option with high operational cost and certain environmental implications. However, it offers some trade-offs because it requires limited land area in comparison with greener technologies and appears to be suitable for urban developments where space is generally a huge constraint.

Treatment performance and compliance with international standards

The treatment performance of the tested technologies is available in Appendix A7. In the UK there are no mandatory standards for greywater recycling against which system compliance can be assessed. This is widely seen as a barrier for its use. To try and provide a fuller picture of the performance of the systems, the effluent quality was compared with international standards and the results are summarised in Table A4.14.

In general, the capital and operating costs for a greywater recycling system vary significantly depending on the scale and extent of treatment employed. The WaND research project developed an economic assessment tool (Memon *et al*, 2005) to capture the interaction and influence of water demand management measures on the whole-life cost of greywater systems (see Chapter 6). The tool application was demonstrated using published data from two example greywater recycling systems. The example systems are briefly described in Box A4.8 and the main costs and benefits shown in Tables A4.17 and A4.18. The analysis indicates that economy of scale has a significant influence on the possible saving.

Future

The application of treated greywater for toilet flushing could result in 20 to 30 per cent reduction in per capita potable water consumption. Although technologies are available for greywater treatment and are widely used in some countries, greywater recycling in the

UK has remained slow compared to rainwater harvesting systems. This is likely due to the energy consumption and costs of these systems when used on a small scale, and the complications inherent with managing any such systems on a community level. The Environment Agency (2008a) advises the use of GWR as a last resort and preference should be given to water-saving household devices. However, large-scale developments should always consider greywater recycling systems. With technological advancement, increased use and acceptability, and a lowering of cost, formal national standards for GWR systems would greatly help their delivery.

Main points on greywater systems:

- the supply of greywater is fairly stable, consistent and broadly matches the demand for toilet flushing. As long as the house is occupied there will be some greywater to reuse
- greywater systems are attractive for buildings with limited catchment area for rainfall relative to the number of occupants (such as apartment blocks and many hotels) and where there are advantages in reducing the overall volume of discharges to the sewer or local sewage treatment
- greywater can also be used in offices to provide mains-water savings of around 20 to 25 per cent
- the capital cost of a greywater system can be lower than that for rainwater systems due to the smaller storage tank but the operating costs per cubic metre of water are potentially higher due to increased maintenance and disinfection costs (MTP, 2007). The main disadvantage of greywater compared to rainwater is the need (in most cases) to provide disinfection, which increases the technical complexity of the system
- scale, household occupancy and water-tariff structure all influence the payback period of greywater systems. If they are deployed at a community level (ie serving multiple residential units), they can be financially viable but the payback period for single-household systems is very high
- the level of greywater treatment required is a function of system scale. For large systems, a multi-stage treatment (eg storage, coarse filtration, biological treatment, fine filtration, disinfection) is required.

4.3 STORMWATER MANAGEMENT

The management of stormwater quality and quantity is an important part of water cycle management. Urbanisation leads to the construction of new roads, surfaces and buildings, which:

- increases impermeable area
- reduces natural percolation of stormwater and infiltration of water into the ground
- disturbs natural drainage paths
- contributes to localised flooding
- reduces the ground water recharge rate (crucial in water-stressed regions)
- contributes to diffuse water pollution.

Stormwater from impervious surfaces may contain pollutants arising from vehicle traffic emission (eg oil and grease). These pollutants, if not intercepted, will eventually drain into receiving waters and can cause damage to aquatic life. A larger impermeable area will increase the surface runoff rate. The increased rate may cause soil erosion and sediment build-up in watercourses.

4.3.1 SUDS

To address these issues, the concept of sustainable drainage systems (SUDS) has developed. This concept involves managing stormwater locally (as close to its source as possible) to mimic natural drainage and encourage its infiltration, retention and passive treatment. The benefits attributed to SUDS include: flood risk management, improved water quality, protection and promotion of natural habitat and biodiversity, groundwater recharge, reduction in soil erosion rate and later sediment build-up rate in watercourses, and creation of recreational features.

It is important to consider the SUDS options that are most appropriate for the site. Local conditions should be considered when specifying the SUDS to be used. Building Regulations specifies that infiltration SUDS should be located at least 5 m away from buildings. This is to stop infiltration flows affecting the foundations of houses and to prevent flooding from groundwater recharge.

In the UK, SUDS are promoted as the preferred drainage option by national planning policies as presented in Table 4.1.

Table 4.1 *National planning policies for SUDS promotion*

UK Country	National Planning Policy
England	PPS25: *Development and flood risk* (CLG, 2006)
Scotland	SPP7: *Planning and flooding* (Scottish Executive, 2004)
Wales	TAN15: *Development and flood risk* (National Assembly for Wales, 2004)
Northern Ireland	PPS15: *Planning and flood risk* (DoENI, 2006)

The main SUDS techniques and their positive and negative attributes are summarised in Table 4.2. Box A4.9 provides a list of important documents that give information on various aspects related to the planning, design, construction and management of SUDS.

Table 4.2 *SUDS performance (source Woods-Ballard et al, 2007)*

Priority	Hydraulic benefits –small events	Hydraulic benefits – large events	Water quality benefits	Amenity benefits	Ecological benefits	Land take	Land availability	Rainwater harvesting benefits	Construction and operation
Water butts/storage	Best	Worst	N/A	Best	Worst	Useful	Useful	Useful	Best
Pervious pavement	Useful	Useful	Useful	Best	Worst	Useful	Useful	Useful	Best
Filter trench	Best	Best	Best	Worst	Worst	Best	Best	Worst	Best
Filter strip	Best	Worst	Worst	Best	Best	Worst	Worst	Worst	Best
Swales	Useful	Best	Useful	Best	Best	Worst	Best	Worst	Best
Ponds	Useful	Useful	Useful	Useful	Useful	Best	Worst	Best	Useful
Linear ponds	Best	Best	Useful	Useful	Best	Best	Best	Worst	Useful
Wetlands	Useful	Best	Useful	Best	Useful	Worst	Worst	Worst	Best
Detention ponds	Useful	Useful	Best	Best	Best	Best	Best	Worst	Useful
Soakaways	Useful	Useful	N/A	Worst	Worst	Useful	Useful	Worst	Useful
Infiltration trenches	Useful	Useful	N/A	Worst	Worst	Useful	Useful	Worst	Useful
Infiltration basins	Useful	Best	N/A	Best	Best	Worst	Worst	Worst	Worst
Green roofs	Useful	Worst	Useful	Best	Best	Useful	Useful	Worst	Best
Bio-retention	Useful	Best	Useful	Useful	Best	Worst	Best	Worst	Best

Key	
Worst	(dark shading)
Useful	(medium shading)
Best	(light shading)

Considering the extensive research already in place, the WaND research project focused on:

- developing decision-support tools to assist the feasibility assessment, design and performance evaluation of stormwater management techniques (Chapter 6)
- monitoring SUDS performance
- modelling stormwater flows and their interaction with soil moisture content
- investigating the effectiveness of rainwater harvesting as a stormwater management option.

The main relevant findings are presented in the following sub-sections of this guidance.

Guidance and evaluation tools for SUDS

National planning policies recognise SUDS as a preferred option for stormwater drainage, so it is important to assess the feasibility and effectiveness before commissioning detailed scheme design. The traditional design and evaluation of the performance of a stormwater drainage system is not based on sustainability, but on hydraulic performance. This is usually associated with the capacity of the system to deal with extreme events (referred to as level of service), rather than evaluation of total performance.

The WaND research project developed an alternative approach for the drainage system performance assessment using both time-series rainfall data and sustainability indicators. It led to the development of several tools that can be broadly classified into two groups:

1. Tools for preliminary drainage design.
2. Tools for the sustainability evaluation of detailed drainage design.

These are briefly described in Chapter 6. Further details on their development and application can be found on the WaND portal (see CD-Rom included with this guidance and Appendix A1). These tools are also available as freeware and can be accessed via <www.uksuds.com>.

The sustainability assessment methodologies developed in WaND were tested at Elvetham Heath (Box 4.2).

Box 4.2 *Stormwater drainage performance sustainability evaluation methodology testing at Elvetham Heath*

Elvetham Heath, in Hampshire, is one of the foremost developments in the country in terms of design concept, where a range of WaND outputs were tested. The development uses some SUDS components to manage stormwater.

The stormwater drainage system evaluation methodologies developed in WaND were tested using an Infoworks CS model of the real-life drainage system at Elvetham Heath. This exercise focused on evaluating the hydraulic and water quality performance from the perspective of sustainability, rather than just the level of service, which only focuses on the protection against sewer flooding.

Present and future rainfall conditions were used with an allowance for climate change to assess the system performance for three different drainage design scenarios. These are:

1 As-built drainage solution containing some SUDS.

2 Traditional piped drainage solution with underground storage.

3 Best practice solution that increases the use of SUDS and includes rainwater harvesting.

Sustainability measures (or indicators) considered for hydraulic performance are flow volume, peak flow rate and infiltration volume. All drainage scenarios were compared with greenfield (ie pre-development) conditions. This analysis was conducted using design storms, individual events from both the five year and 100 year rainfall series and a contentious five year rainfall series.

The measure of sustainability for water quality consists of a qualitative evaluation of treatment provided by the drainage system. This is based on categorising land-use (ie residential, commercial or industrial) and the pollution mitigation effects of the drainage system used, including different types of SUDS techniques (eg detention basins, green roofs, filter strips, infiltration trenches, pervious pavements, oil interceptors, ponds, soakaways, swales and wetlands).

Figure 4.1 *Detention ponds at Elvetham Heath*

Figure 4.2 *Swale at Evetham Heath*

Figure 4.3 *Pond at Elvetham Heath*

The analysis showed that a good sustainable solution for a drainage system based on a measure of the greenfield behaviour of the site is difficult to achieve and highlight the importance of maximising the benefits available from the range of SUDS components

SUDS field monitoring

SUDS promotion is largely based on their theoretical and short-term analytical performance in terms of pollution abatement and flow attenuation. SUDS are advocated by the Government, the Environment Agency and local planning authorities. There is a need to build the evidence based on the long-term performance of SUDS. To address this, a SUDS field monitoring programme was initiated to:

- improve the understanding of SUDS performance at the development scale with respect to drainage, water quality and the effect on surface and groundwater
- investigate SUDS processes and response times for different SUDS combinations (individually and in trains), different timescales (including over annual cycles) and implications on receiving waters
- develop a continuous simulation of soil and runoff processes incorporating SUDS, using monitored data and scenario modelling.

A programme was established to monitor 14 main sites in four developments served by SUDS (Elvetham Heath, Banbury, Witney, and Cowley). Two further case studies were considered. The first and more detailed covered two sets of pond/wetlands in Pitty and Chellow Dene Becks, Bradford, built as trial "flow conditioners" to treat urban runoff to a standard suitable for water features. The second was a qualitative study of the performance of new wetlands/lakes in a housing area in Sheffield. On these sites, sophisticated instrumentation was deployed to monitor rainfall, runoff, soil moisture and variations in water quality. The preliminary observations made from the ongoing monitoring programme are presented in Box 4.3.

Box 4.3 *SUDS field monitoring – preliminary observations*

- the SUDS monitored worked effectively, but did not perform exactly as planned due to different construction, poor maintenance, unexpected flow paths, overlooked functions and overflow paths
- soil moisture content was found to vary greatly across the urban areas studied, affecting the spatial distribution of pervious area runoff. However, local moisture content can be predicted using daily rainfall history and a local value of maximum soil water capacity
- dry weather sampling at the Bradford ponds found that the proposed aesthetic standards used so far have been met
- spot sampling for faecal coliforms showed that removal was greatest during sunny periods with longer exposure time achieving concentrations comparable to a UV-treated effluent
- the Sheffield wetland study found that over a year their aesthetic quality remained high and vandalism was generally low. Clogging of outlets was a problem, although this was linked to the construction phase.

Costs

Duffy *et al* (2008) have identified the Dunfermline Eastern Expansion Area (DEX) as probably the only UK site where detailed data is available to determine the real cost of SUDS construction and maintenance in comparison with conventional drainage systems (Box 4.4). The cost analysis is supportive of SUDS and indicates that well designed and maintained SUDS are more cost effective to construct, and cost less to maintain than traditional drainage solutions, which are typically unable to meet the environmental requirements of current legislation.

Box 4.4 *SUDS implementation in the Dunfermline Eastern Expansion Area (DEX)*

Site description

The Dunfermline Eastern Expansion Area (DEX) is a 550 ha (5.5 km^2) site that lies to the east of Dunfermline. The site, which was predominantly greenfield, will be developed over the next 20 years as a mixture of industrial, commercial, residential and recreational areas.

The Scottish Environment Protection Agency (SEPA) is primarily concerned with the quality of the runoff compromising the downstream river quality. However, the catchment immediately downstream of the site has an existing flooding problem, and SEPA are also concerned about the velocity and volume of runoff from the site. For these reasons, using SUDS became a planning condition, and DEX is currently the largest site in the UK to use widespread sustainable drainage methods, with the regional framework system being put in place before development started.

Main SUDS used

DEX is located on an area of predominately low-permeability clay soil. Infiltration methods have limited application on the site. Some residential roads will be served by soakaways, where soil permeability permits. Much of the spine-road system is drained using offlet kerbs, filter drains and swales, which discharge into extended detention basins and wetlands that also serve adjoining housing areas.

Figure 4.4 *Wetland area*

Figure 4.5 *Detention basin*

Treatment of surface water runoff from the development and roads is achieved through a system of regional ponds and wetlands before discharged to the watercourses. Ponds and basins are widely used to achieve maximum attenuation of storm flows. The wetland is located in a public park area where informal public open space adjoins an existing forested area and an area set aside for football pitches, a rugby pitch and tennis courts. Permeable paving has been used in the supermarket car park, which is connected through attenuation/infiltration basins to the wetland.

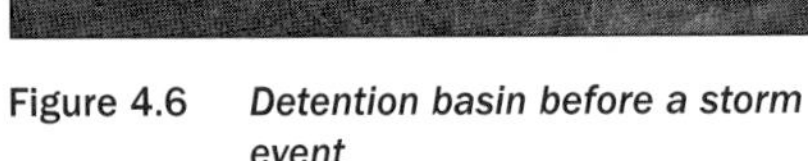

Figure 4.6 *Detention basin before a storm event*

Figure 4.7 *Detention basin after a storm event*

Benefits and barriers

Benefit: the use of conventional drainage systems would have resulted in prohibitive costs to drain the development (the discharge being required via a 5 km sewer to the Forth River). The use of SUDS was promoted by consultants and the developer alike as a means of achieving an economic solution, which was taken up by the planners.

Barrier: adoption. The highways authority was initially unwilling to adopt SUDS for road drainage. But now it has adopted the SUDS on principal roads.

Other issues

The local councillors were anxious about safety near open water. However, barrier planting and shallow reed-planted margins have removed this concern.

Environmental impact of SUDS

In terms of the effect Voaden, (2008) performed an outline assessment for a range of SUDS including swales, filter drains, retention and detention ponds and compared their performance with conventional drainage systems. A range of damage assessment categories (life cycle assessment) were used including human health, ecosystem quality and resources:

- **human health:** how the different methods of water management will affect the wellbeing of humans. This includes any negative effects to their physical well-being through disease, injury or infection.
- **ecosystem quality:** the effect of delivering water management options in relation to the biodiversity of the natural environment.
- **resources:** the number and amount of resources used to supply and build the various water drainage options.

The results are summarised in Figures A4.7 and A4.8. Figure A4.7 suggests that relatively filter drains have the poorest environmental performance. However, Figure A4.8 suggests that SUDS have significantly lower environmental impact compared to conventional drainage scheme.

4.3.2 RWH as a stormwater management option

The effectiveness of RWH systems as a demand management option is fairly established. However, the information on the quantification of benefits that RWH systems could offer as a stormwater management option is limited. The size of the rainwater collection tank plays an important role to yield optimum benefits. For RWH to be effective as a water demand management option, it is desirable to have the tank fully filled with rainwater. However, to best attenuate stormwater flows, an empty collection tank is desirable. These conflicting factors need a trade-off to partially satisfy both goals.

The WaND research investigated the influence of rainwater storage provision, and its use for non potable application on surface runoff volume reduction see Appendix A4.7.

Future

National policies require considering SUDS as a preferred option and surface runoff rates to remain broadly unchanged compared with the pre-development situation. The EU Water Framework Directive also implies delivering catchment-based drainage approaches including SUDS. Delivery of SUDS is likely to increase particularly with the introduction of the Flood and Water Management Bill (2009a) see Chapter 2. However, in the interim there are some barriers to overcome (see Chapter 5). These include addressing SUDS adoption issues and putting in place safeguards to discourage vandalism.

4.4 CONCLUSIONS

Water-saving devices

- installing water-saving devices in new housing may only slow the overall increase in per capita consumption in the medium term, but can have a significant affect in the longer term, especially if devices are retrofitted into existing housing stock

- the major demand moderating factors are metering, toilets, normal-flow showers and modern efficient white goods. Improved plumbing and heating will also contribute
- relatively cheap and simple water conservation devices, such as low-flow taps, aerated showers, low-flush toilets and simple rainwater butts, can offer short payback periods, use less energy and involve less maintenance, so should be considered before rainwater harvesting (RWH) or recycling of greywater
- innovative devices are emerging and some can provide considerable water savings (eg the ULFT), with comparable levels of service to other devices.

Stormwater management and rainwater harvesting

- although SUDS are expected to contribute towards attenuating peak flows and providing runoff treatment, fully recovering the greenfield hydrological response has not yet been proven with the examples studied
- SUDS schemes have been shown to have a significantly lower environmental effect than conventional drainage, and can cost less to construct and maintain
- the use of infiltration SUDS in new developments should be exercised with caution because of the interaction between infiltrated flows and foundations of houses and to prevent flooding associated with changes in groundwater levels
- stormwater generated pollutant loads and SUDS treatment performance is extremely variable, indicating that the performance of SUDS units varies greatly depending on the site, the construction methods and maintenance regime
- RWH systems have the potential to manage stormwater peak flows. Retrofitting of urban areas with rainwater reuse systems, where flooding has historically been relatively frequent, could reduce the flooding frequency and flooding volume
- the effectiveness of RWH systems as a water demand management option is established and they are widely used in many parts of the world. For a cost-effective system, the main issue is establishing the optimum size of the tank keeping in view the water demand for non-potable applications (eg toilet flushing). Tools are available to quantify the effect of tank size on overall water savings
- for RWH systems, the major cost is the storage tank and pump unit. The roof area of a typical single household is not sufficient to meet non-potable water needs fully. The expense of a two or four- house system is not much more than a single-house system. The only problem is arranging for the equitable sharing of water benefits and maintenance costs
- the rainwater use in affordable low-rise housing is constrained by the roof area. Only larger or less densely occupied private housing is likely to achieve the 50 per cent mains-water savings, unless surface level catchment is also considered
- where there is insufficient rainwater to supply all non-potable uses, distribution can be limited to one outlet, which minimises pipework costs.

Greywater reuse

- greywater systems are attractive for buildings with limited rainfall catchment area relative to the number of occupants (such as apartment blocks and many hotels) and where there are advantages in reducing the overall volume of discharges to the sewer or local sewage treatment
- greywater can also be used in offices to provide mains-water savings of around 20 to 25 per cent

- in comparison to RWH systems, greywater recycling systems offer a reliable supply for non-potable uses. Economy-of-scale benefits could be realised in cluster or neighbourhood-scale systems. However, these are not "fit-and-forget" systems and require regular maintenance
- in social housing, the residents do not benefit from savings offered by the communal systems, since housing associations are responsible for their water bills. Housing associations could be direct beneficiaries if they install and manage greywater recycling systems centrally
- relatively high installation and maintenance costs, unavailability of any formal guidance and reuse standards are perceived as the main barriers to taking up greywater recycling systems at the right scale
- although natural greywater treatment technologies (eg reed beds) have limited environmental effect, they have high land requirement and are susceptible to shock loadings. Energy-intensive technologies (eg membrane bioreactors) have high operational cost and a bigger environmental footprint, but they have high resilience to shock loadings, limited space requirement and are potentially suitable for urban developments.

The technologies available for water cycle management vary in scale, cost and complexity. The health risks of well maintained SWCM technologies are generally low. These benefits should be considered when comparing the cost and complexity of the technologies. Water supply management and stormwater management are the main focus points on achieving sustainable water management in new developments.

It is generally agreed that the primary methods of achieving sustainable water management in new developments is through water-saving devices and SUDS. Water saving devices are relatively easy to install, can considerably reduce water consumption and have noticeable short payback periods. Their inclusion in new developments should be increasingly common and may become mandatory as environmental legislation develops. SUDS are the main focus in stormwater management and are being championed within planning policy. Finally GWS and RWH can help obtain high levels of sustainability but have greater delivery and maintenance costs. They are most beneficial on a larger scale.

4.5 FURTHER RESEARCH AND GUIDANCE

- further research is needed on the cost, reliability, operation, maintenance and payback of GWS and RWH
- more specific guidance is required on the technologies and their application, for example, tank sizes or exact costs
- further research into the performance, costs and benefits of SUDS is required
- more independent research on technologies not covered by this research is needed
- the evaluation of the sustainable drainage benefits of rainwater harvesting.

5 Stakeholder engagement

This chapter:

- identifies the factors that influence costs, benefits and risks to stakeholders (ie responsible bodies and end users including the general public), and the social issues and barriers to using a SWCM
- highlights opportunities to improve stakeholders engagement
- identifies principles for achieving effective stakeholder engagement and presents ways to evaluate and frame important messages.

5.1 INTRODUCTION

Conventional water cycle management on new developments requires little multi-stakeholder engagement, and is largely undertaken by the water utility and developer. However, delivering SWCM is a relatively novel concept, requiring a greater degree of creativity and interactive problem solving, and with engagement of a wide range of stakeholders, particularly the local community. This chapter sets out an approach for involving stakeholders based on case study evidence.

5.2 LEARNING THROUGH CASE STUDIES

The research shows that successfully delivery of SWCM will lead to many innovative processes that will be unfamiliar both to bodies responsible for managing the water cycle and to the general public. It is widely accepted that SWCM can be achieved through a sustained and managed process of stakeholder engagement.

This chapter sets out the issues involved in stakeholder engagement and describes how this can be successfully achieved. It also highlights important information by referring to the three case studies that formed part of the research programme. These are:

1. A sustainable drainage scheme in Sheffield, South Yorkshire.
2. A domestic rain and greywater recycling scheme in Childwall.
3. Sustainable drainage provision in Elvetham Heath, Hampshire.

The case studies were chosen to represent the widest possible range of stakeholder issues that might arise during the delivery of SWCM. However, the case studies do not cover the full range of potential SCWM strategies. There is also enough evidence to identify the main issues that arise from stakeholder engagement and from dealing with a wide number of responsible stakeholder bodies.

5.3 DESCRIPTIONS OF THE CASE STUDIES

A brief description of these case studies is given in Boxes 5.1 to 5.3 and illustrated in Figures 5.1 to 5.6. Further details of the case study characteristics, research methodology employed and analysis is provided in Sharp (2007) and this paper can be accessed via the WaND portal (Appendix A1).

Box 5.1 *Sheffield case study - SUDS*

SUDS were located in a new district park in Sheffield to handle roof and road drainage from a new mixed social and private housing development. The SUDS were initiated through a proactive local NGO. The case study is important as it is a retrofit application of SUDS in an existing inner city area. Local residents were unaware of the concept of SUDS at the start of the project. There was a history of vandalism and anti-social behaviour in the area so any new surface development needed to address this problem.

An individual "champion" played a vital role in developing the scheme, with considerable assistance from the local council who agreed to adopt the SUDS for a commuted sum from the developer. As the project developed, ongoing community education and engagement was seen as an important means of preventing vandalism and developing a better community understanding of sustainability. It was found that SUDS represented an "all-win" scenario with benefits for a wide range of stakeholders including the developer, the local authority, the water company and the local community.

Figure 5.1 *SUDS from Sheffield case study*

Figure 5.2 *Wetland from Sheffield case study*

Box 5.2 *Childwall, Liverpool case study – water saving devices, rainwater harvesting and greywater recycling*

Sustainable water features comprising low-flow taps and dual-flush toilets formed part of an environmental demonstration project adopted in social rented accommodation in Childwall, comprising 34 houses and an elderly care home. All properties in the scheme were fitted with these features. Also, 15 properties and the elderly care home were fitted with a communal rainwater system and 12 properties were fitted with a hybrid rainwater/greywater recycling system.

Stakeholder commitment and funding were important to the development of the project. The project helped a range of stakeholders develop expertise in sustainable technologies. However, considerable technical difficulties were experienced, at a cost both to the tenants and to the social landlords. These difficulties were associated with a lack of engagement involvement with the tenants throughout the process. This would have encouraged a greater awareness of sustainable issues and how the water technologies worked including a better buy-in to the project from participants.

Figure 5.3 *Communal rainwater recycling system pump room at Childwall (courtesy Steven Kennedy)*

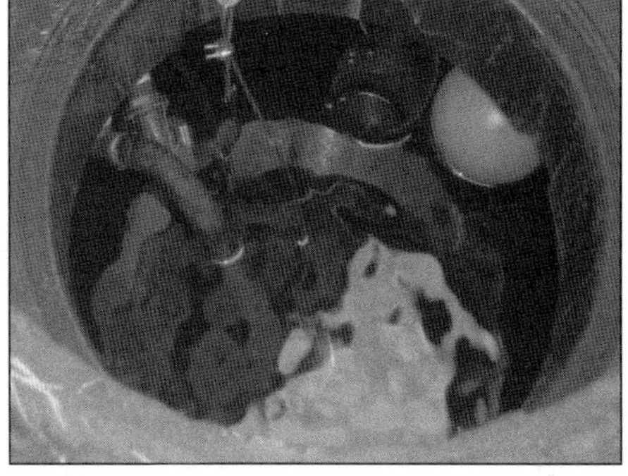

Figure 5.4 *Individual household rain and greywater recycling underground tanks from Childwall (courtesy Steven Kennedy)*

Box 5.3 *Elvetham Heath case study – water saving devices and SUDS*

SUDS and water-saving fixtures (dual-flush toilets and water-efficient washing machines) were installed in a very large development involving multiple housing developers in Elvetham Heath.

The SUDS element of the scheme proved particularly controversial to the stakeholders and only emerged following prolonged negotiations between the developers and the local council. The scheme could be seen as offering multiple wins to the developer, householders and council. In particular, the SUDS scheme could fulfil requirements for open space on the development, while also performing a drainage function and providing considerable aesthetic appeal. There appears to be general acceptance of the water-saving fixtures. However, the use and effect on water savings requires continued monitoring and analysis. Figures 5.5 and 5.6 demonstrate some of the SUDS systems used at Elvetham Heath.

Figure 5.5 *SUDS at Elvetham Heath*

Figure 5.6 *Pond at Elvetham Heath*

5.4 PRACTICAL DELIVERY OF SWCM

Although the long-term objective is to develop sustainable homes in the UK, the research has demonstrated various issues to delivering SWCM at the present time. With any new methods or technology the associated problems occur when people are unfamiliar with them. This is a natural response but the benefits are generally indisputable for SWCM. The main concerns are financial, legislative and cultural and were demonstrated in different ways in all three case studies.

If SWCM is to become normal practice in new developments there needs to be a change from present policy and practice. Now significant progress is possible, as the case studies demonstrate, especially where the early lessons learnt influence its delivery.

5.4.1 Planning policy and strategic implementation

In all three case studies, successful delivery of SWCM required underlying policy, and the early engagement of stakeholders, both professional and end users, so the following should be considered:

- **incorporating management plans into local development frameworks:** This is needed to enable good practice to develop over time. In the Sheffield case study a lack of legislative pressure to adopt a SUDS scheme was noted by many stakeholders. Although planning policy such as PPS 25 advocates SUDS as the preferred way of dealing with surface water runoff, there is currently no requirement to conform to this advice either in planning law or Building Regulations. The Flood and Water Management Bill 2009 proposes that new developments will no longer have the automatic right to connect to the sewers for surface water runoff. Instead SUDS should be used where possible.

  At Elvetham Heath the lack of a co-ordinated approach was cited as a barrier to its adoption. Traditional approaches, policy and practice tend to be uniform across England and Wales, but with SWCM different approaches can appear in different regions. This causes problems with developers and other stakeholders working across a wide area.

  In Sheffield there was consensus that adopting SWCM initiatives was not a priority and that local authorities are now acting in isolation from one another rather than assisting each other with this issue (such as through networks and agreed guidance documents). Although some efforts are being made, it is suggested that knowledge of adopting SWCM initiatives should be better co-ordinated across government departments and agencies.

- **identification of a lead body:** as shown in the Sheffield case study, a champion to lead the delivery of SWCM can prove very effective. Local authorities can play an important role in the initiation of SWCM schemes in many: The Flood and Water Management Bill 2009 propose that local authorities take the lead role in local flood risk management:
  - as the local planning authority, their knowledge and awareness of sustainable water management is vital. The local planning authority holds the power to grant planning permission for development and can decline planning applications on the basis that SWCM has not been given due consideration
  - in cases where the council is selling land to housing developers, they are able to attach certain conditions to the sale. One such condition could be that the local planning authorities include policies that require the incorporation of SWCM in new developments unless the developer can demonstrate that it is impracticable.

This means that developers have to proactively adopt SWCM, rather than voluntarily including it in planning applications

- the local authority, as was the case in the Sheffield project, could opt to use their power to adopt and maintain certain elements of SWCM schemes, eg SUDS. SUDS schemes meet several objectives for local authorities concerning new developments such as the creation of a better environment, promoting nature conservation, improving wildlife habitats and the provision of sufficient open space to meet residents' needs. By offering to adopt SUDS, in exchange for a suitable commuted sum from developers, local authorities would remove the barrier to their adoption. Indeed the Flood and Water Management Bill 2009 proposes that county or unitary councils, where relevant, should adopt new SUDS that have been constructed in accordance with national standards. The local authority can then choose to conduct the long-term maintenance themselves or to use a contractor.
- until SWCM is made explicit in policy and later integrated into legislation, it will always be dependent on voluntary initiatives, or the requirements of each individual local authority.

- **early involvement of stakeholders:** this should include the various bodies responsible for the water cycle, the developer and the end user (eg the local community). Where a water cycle study has been carried out at a strategic planning level, the institutional stakeholders will already be engaged in SWCM, and there are likely to be fewer barriers to achieving SWCM.

  Observations made at the Childwall site indicate stakeholders' lack of commitment as one of the main reasons for the failure of the hybrid systems (ie greywater and rainwater systems). Feedback from some stakeholders suggests that making sure all stakeholders are fully signed-up to the process from the start would help to encourage a sense of commitment.

  The different needs of different stakeholders should be considered. For example, at Elvetham Heath there appeared to be strong resistance by developers and owners to the delivery of SWCM options, except for dual-flush toilets and water-saving washing machines. Developers are reluctant to adopt such measures unless required to do so by law (such as through a change in planning and/or Building Regulations).

  From the developers' perspective, it appears that:

  - building eco-friendly houses is risky, with the increased cost deterring demand and reducing profits
  - the majority of property buyers and tenants are more interested in schools, shops and open amenity, rather than the availability of rainwater collection devices
  - there is a potential health risk posed by some water-saving systems.

  It follows from the above that effective and timely engagement of stakeholders is vital for the successful delivery of SWCM. This should be formally planned into the delivery process. Engagement with the public can be problematic, with responsible bodies needing further guidance on how to progress this effectively. Section 5.6 is devoted to offering guidance on effective public engagement.

- **accounting for different responsibilities:** different stakeholders have different responsibilities in the management of the water cycle and it is important that these are clearly defined, appreciated and managed. As an example, the Sheffield case study showed that SUDS are multi-disciplinary with several stakeholder groups holding the potential to initiate or assist the development of schemes. It is important that the responsibility for SUDS initiation is clearly assigned and accepted.

5.4.2 Incentives

The cost of water services is relatively low compared with other utility services (such as energy).Using financial incentives to promote participation may not be the most effective strategy. This was particularly true in the Childwall case study where end users tended to be encouraged by the financial aspect rather than any environmental improvement. It was found that reducing water bills to encourage residents to try new technologies can offer a strong incentive, but the effect is temporary. The financial incentive increases expectation, and if end users do not experience any financial benefit, they can feel cheated. The extent of money saving should not be exaggerated and should be sustainable.

Water is still seen as a commodity and the value is such that wasting water is not considered punitive by the purchaser. Even where water meters are installed in all households, users tend not to read them or monitor their consumption.

5.4.3 Reliability of new systems

Conventional water and wastewater services have evolved over many years. Although some important developments have arisen from structured research and development investment, many have been ad hoc and in response to observed deficiencies in previous systems. Such long-term development delivers good reliability and a strong familiarity with installers and end users.

Where innovative systems are adopted, reliability can often be questionable and many installers may not be familiar with the different requirements of new systems. Equally, end users may not understand the different operational constraints and/or they may not match their lifestyles. As installers and users will tend to revert to traditional practice this can lead to innovative systems not functioning as intended, and this leads to poor performance and a lack of confidence. Because new technologies impose different requirements on installers and maintenance contractors, structured training should be included in SWCM plans. This can help to avoid, or promptly identify and fix malfunctioning equipment.

Overcoming these difficulties should not be underestimated. The need for proving new techniques is vital and this is why case studies and pilot projects are important. Great care is needed when planning SWCM measures. In Childwall it was estimated that a period of one to three months was required for planning and pre-testing water-saving systems. Simple, low-maintenance technologies were preferable to more complex systems that required technical support.

As SWCM techniques and technologies become more widely used and evolve over time, as traditional infrastructure has, there will be less uncertainty about their reliability, operation and cost, and increased use even from the "less adventurous" practitioner.

5.4.4 Acceptance of SWCM measures

SWCM can deliver substantial environmental and financial benefits so many of those responsible for water supply, drainage and wastewater disposal, and for urban development are justifiably enthusiastic.

In practice there is little knowledge by those who deliver schemes (eg developers) or by end users (eg the public) on the benefits of SWCM. As the case studies show, this difference of view can arise because the benefits may not align with the needs or aspirations of stakeholders, or because apparent benefits are not realised in practice.

The case studies indicate that careful management is needed if the benefits of SWCM are to be realised. In general, end users do not get much opportunity to be involved in the selection of SWCM measures. Excluding end users, such as owners and tenants, from the process of choosing the SWCM technologies can result in undesirable consequences. This is particularly important when retrofitting measures. In Childwall, failure to keep the selection process transparent, exchange information openly and promote active participation led to negative feedback where problems arose. A strong sense of ownership and control needs to be fostered at every stage of the process.

The Childwall case study also showed that before new SWCM solutions are delivered, it is crucial to consider the potential users' characteristics, social background and cultural practices, since all these factors may affect the performance and acceptability of SWCM strategies. Monitoring the performance of SWCM measures is important in terms of getting stakeholder buy-in and demonstrating wherever possible financial as well as environmental benefits. The financial benefits may do more to help stakeholders accept new technologies than environmental benefits.

5.5 EFFECTIVE PUBLIC ENGAGEMENT

5.5.1 The need for engagement

The need for public engagement has been demonstrated in the case studies in Section 5.3. This need arises because:

- some organisations have to engage the public with respect to SWCM because it is obligatory, for example, water-service providers with respect to water efficiency
- many elements of SWCM move beyond technical solutions, and require some change in public behaviour. These changes cannot be forced, but need to be delivered through incentives or persuasion. These changes are more likely to happen if they match the context of users' lives
- even where solely technical fixes are employed, they should consider the effect on the public (Section 5.3). Engaging the public can help to ensure that innovation works effectively
- in practice, many organisations may be communicating a particular perspective on SWCM (and of the public) in their usual communications, for example, through water bills, advice, or policies (Sharp, 2006).

To achieve effective public ownership of SWCM, it is important to understand their perceptions and experiences on water management first, before designing and adopting any communication strategy.

5.5.2 Understanding end user behaviours and perceptions

To help understand behaviours and perceptions, the WaND research analysed responses collected from the case studies in Section 5.3. These were supplemented by two further studies, conducted specifically to investigate users' perceptions and SWCM promotion (Box 5.4). These findings are summarised as:

- when asked about water, respondents' immediate associations come from media reports, and they usually concern floods and droughts, which presents a negative picture of water management
- in contrast, the public's everyday experiences of water remain undisclosed, but when asked about water in their daily life, they are very positive

- respondents were confused about how water in the environment relates to water in the home. This was demonstrated by their disbelief that the UK can be short of water after recent rainfall and their misunderstandings about the mechanisms for managing water, including confusion between energy and water-saving actions
- respondents normally talked about water in the environment as nice "elsewhere" (for example, on holiday), but would make negative associations with external water in their immediate area
- when discussing their own use of water, respondents chose particular patterns of behaviour according to specific pressures on their time and their perception of being non-wasteful
- in contrast, when discussing others' use of water, there is an assumption that people will always be wasteful, and that people will only save water if encouraged by pricing and metering incentives
- respondents' patterns of water use are also taken in a context of distrust of those in control – whether that is the Government, water companies, or local authorities.

So, the analysis identified some important issues that need addressing when getting the public involved and generally people:

- do not think that there is an issue with water in the UK
- have a belief that water is plentiful
- have distrust in authority and their communications, particularly about water
- have a lack of knowledge of water processes.

These underlying perceptions make it more difficult to achieve changes in water use behaviour. The issues can be addressed on a local scale, particularly if ways are found to connect to people's local experiences of using water. But achieving widespread behavioural change may not be possible unless there is a concerted effort among influential organisations to initiate a wider cultural change to help society understand the existence (and importance) of the water cycle and their part in it.

Whether national or local, all water-related communications can build on:

- the positive experiences of water in the home and external water elsewhere
- identifying how to conserve water
- ideas on how fair metering is
- emphasis of accepted ideas on energy savings and the connection with water.

Local examples of water and the environment should be used to address people's lack of understanding about the whole water cycle. Developing and communicating a set of coherent messages can only be done through interaction and engagement of all stakeholders.

Box 5.4 *Supplementary study on user interactions with SWCM strategies*

In an independent action-based research study, the WaND project attempted to identify how user behaviour interacts with water using technologies with the aim of exploring changing (and unchanging) perceptions of domestic water use with respect to water scarcity. A case study was initiated involving a new housing estate in Essex where water-saving devices were installed and water savings monitored between 1998 and 1999.

Householders were interviewed before and after the efficiency campaign. The features of the case study and the research methodology employed are reported in Knamiller *et al* (2007). The findings and recommendations are summarised as:

Findings

- many householders do not understand how, or do not believe, water supplied to households can be a scarce resource. There is a perception that there is high rainfall in the UK, so the link between domestic water and environmental water is confused
- people are likely to change originally installed water-saving fittings to the ones suiting their lifestyles
- many of people's everyday actions with water are performed through habit. Deliberative decisions are not made about how an action is performed as it is done automatically, which is why habits are notoriously difficult to change (Jackson, 2005)
- people develop water use habits to make life easier and fit around their needs and routines. Time constraints are a common factor and health issues also influence behaviour
- metering appears to have a variable affect, with some of those interviewed feeling it made them more conscious of their consumption and others feeling that they forgot it after a while, or could do nothing about reducing their use anyway. Also, often only one member of the household pays the bills, usually by direct debit so appeared unaware of their bills. Metering has little effect on water use in these circumstances
- water management learnt via the media is negative. People have a negative perception of the water companies in terms of leakages and costs. There was a common notion that water companies were doing little to address water shortage issues, and some interviewees questioned why they should be making small savings when water companies were not managing the systems
- the positive personal experience of everyday water is implicit.

Recommendations

- the diversity and variation in water use suggests that blanket fitting of the latest water-efficiency devices may not be helpful. Rather, a more adaptable form of technology promotion would seem more desirable. Households could be offered a choice of water-saving equipment appropriate to their circumstances
- to counter the negative perception of disastrous water management reported by the media, we need to encourage discussions about people's positive everyday experiences of managed water. This could be achieved by:
  - making the connections between domestic and environmental water more visible
  - civic groups may also play significant roles, as might schools, workplaces, leisure centres, local radio etc by broadening the contexts and storylines to connect people with water.
- however these will only be effective if the activities address the day-to-day issues that really concern people in locally relevant ways
- water saving might be more effectively promoted through other pathways. For example, the role of plumbers as significant intermediaries has been given minimal consideration within demand management strategies.

Further details of the study and explanation of findings and recommendations is available from the WaND portal (Appendix A1) and Sefton and Sharp (2007).

5.5.3 Framing effective messages

The WaND research conducted an extensive literature search to identify factors influencing the reception of messages aimed at bringing pro-environmental behavioural change (Sefton and Sharp, 2005). These factors were then translated to develop a framework for effective SWCM messages. The framework is summarised in Table 5.1.

Table 5.1 *Checklist for framing effective messages*

Recipient process	Questions	Potential issues
Recognise problem	How do messages present SWCM (eg water efficiency) issues?	• is fear used as a motivator? • are problems framed in a local or global way? • to what extent are messages legitimised through experts or through local knowledge? • does the way the issue is presented support or undermine source credibility? • are incentives used to motivate action, and if so, do they offer ongoing feedback?
Recognise potential to take action	How do messages present the actions required?	• are actions relatively easy to do? • are messages positive (encouraging good actions) or negative? • do actions involve changing habits or behaviour? • could messages be read as engendering guilt or blaming recipients? • what descriptive social norms are implied about people who take water saving actions?
Changing action will address problem	What evidence is given that the suggested actions will address the problem?	• have issues been framed in big way, which could undermine feelings of efficacy? • have issues been presented in ways that develop unhelpful descriptive social norms?

5.5.4 Guidance for good quality engagement at a local scale

Neighbourhood or development-scale two-way involvement is crucial to the development of ownership of SWCM solutions. The emphasis should be on two-way interaction (eg service provider and end user) rather than just provision of information (eg leaflets), so the aim is to:

- make the water (and other) aspects of the (re)development work better for wider stakeholders (eg developers, regulators, service providers)
- ensure that the (re)development also achieves the best possible ends for the people who live, work and play in it.

The factors to consider for effective engagement include:

- plan for and allow resources for engagement processes – change will be more successful by working together, so invest early
- start early, ie by thinking about who will live in or be affected by the changes or the process of change
- work out stakeholders' constraints – what need to be done, and what is flexible according to people's needs

- discuss with local community leaders the best way to engage locally – it might be through a public meeting for the specific purpose, or by visiting multiple community venues during existing meetings
- work out what people's main concerns are, in general, before asking them about the specific information needed to guide your development process
- provide multiple non-verbal means for people to communicate. Don't expect them all to be comfortable with a big public-meeting format and use small groups, maps and diagrams. Overall, there is a need to make the process of engagement fun
- thank people at the end for their input, demonstrating how much it is valued
- communicate with people after having met them about what is the benefit of the process and how it will affect the development
- maintain frequent (regular) contact to make sure that the messages communicated have an effect.

5.6 TAKING UP SWCM IN THE FUTURE

The research in general and the case studies in particular, have shown that there is local enthusiasm and strategic intent to adopt SWCM in future development. Certain methods are necessary for SWCM to become normal practice and not just reliant on voluntary measures.

5.6.1 Organisational methods required

Necessary requirements for taking up SWCM in England and Wales have been identified. These include:

- improve government leadership in defining SWCM and explaining why it is needed, either to planners and developers or to the public at large. This has been addressed to some extent with *Future water* (Defra, 2008a) and the Code for Sustainable Homes, but much more needs to be done
- encouraging regulatory pressure to deliver SWCM
- building stronger linkages between SWCM and other sustainable building drivers such as energy management, which may have higher priority
- central government resolving issues on maintenance and ownership, otherwise local authorities are likely to be hesitant about promoting their adoption. This is being addressed in the Flood and Water Management Bill 2009.

The research concluded that while water increasingly features on the political agenda, mechanisms for integrated and sustainable management of water should be encouraged and given high priority. There are few examples of SWCM innovation in practice. Each of the case studies investigated in the WaND research programme developed "sustainability" expertise for the parties involved. One promoted local learning and a second was part of a national education programme. Further case studies are needed to promote SWCM across the UK, and to encourage their co-ordination and promotion. Also, greater accountability and funding is needed for the development and delivery of water management policy, and raising greater awareness is required to link the public to a role of water stewardship, rather than simply consumers.

As set out earlier, one way of effectively embracing new ideas is through advocates or champions. A proven method of delivering this is through the development of support networks that can also encourage demonstration or pilot projects. There is a need to develop and manage networks such as LANDF♦RM (Local Authority Network on

Drainage and Flood Risk Management) to share SWCM experiences and information and engage with key stakeholders.

5.6.2 Recommendations for taking up SWCM

Based on the evidence collected from the case studies including the questionnaires on water awareness, later analysis and a broad literature search, the WaND project developed a set of five recommendations for change to help improved SWCM (Table 5.2).

Table 5.2 *Recommendations for SWCM uptake*

1	**Mandatory requirements for new housing to be sustainable appropriate to the local and regional context** • a universal mandate applicable across the UK to deliver a "level playing field", based on the Code for Sustainable Homes • the system should also allow room for regional circumstances to dictate how the mandate is fulfilled, for example, water-scarce areas should be able to mandate that all new housing achieve specific standards in relation to water fixtures and fittings. such requirements could be appropriately developed by regional development agencies or local authorities within the existing development planning system • planners and developers should also be encouraged to make judgements on the specific innovations that suit local circumstances by using an appropriate sustainability framework. However, such flexibility should not allow planning authorities to opt out of SWCM practice • the level of the overall mandate should be reviewed regularly, and should increase in-line with developing technologies and changing needs with respect to sustainability • the mandate needs to encompass not just local measures (eg the immediate site), but also area and regional measures (eg drainage relating to new estates).
	Basis of recommendation: it was widely agreed by the case study respondents that for sustainable water management to be more widely adopted there needs to be a mandatory requirement to be included in new homes. The Flood and Water Management Bill 2009 requires developers to demonstrate that they have met national standards for SUDS before they can connect any residual surface water drainage to a public sewer. However, further sustainable water measures such as water efficiency devices in households should also be specified in government policy. The success of such a requirement was demonstrated by the Housing Corporation's expectations with respect to social housing; but similar expectations need to be transferred to all new homes.
2	**Modest financial support can boost learning and innovation in relation to sustainable water management** • in regions or localities where there is a perceived need for water-related innovation, funds to be made available to support professionals in making educative visits to innovative water management locations abroad and in the UK • funds should also support pilot projects that lead the adoption of a new technology. Similar to the European Regional Development Fund (ERDF) funding in the Childwall case study, such support should specify that completed projects open up to other professionals to increase the overall level of learning • courses promoting SWCM and innovation should be an integral part of continuing professional development for planners, developers and consultants.
	Basis of recommendation: interviews for all case studies revealed that there is a barrier to innovation in terms of professionals such as developers, planners, sustainability professionals, site contractors and plumbers gaining practical knowledge about innovative water systems. This lack of knowledge not only made innovation more difficult (for example, in the Sheffield case study), it also meant that critical communications (for example, with residents in the Childwall case study) were not carried out.

Table 5.2 (contd) *Recommendations for SWCM uptake*

3	**There is a need to increase the general awareness of water issues in the UK** • raising general awareness about water management issues would increase the chance of innovations being delivered. Measures to raise awareness might include: ○ greater promotion and availability of sustainable water management devices, such as low-flow taps, water butts etc to enable people to understand how to get involved ○ wider public participation in water-related choices, for example, about whether specific measures should be adopted in a local new development ○ integration of communications about water management with other aspects of sustainable citizenship – few citizens differentiate between these aspects and current systems that divide the promotion of water and energy mean unnecessary and expensive replication of efforts.
	Basis of recommendation: the delivery of sustainable water management in the UK faces an uphill battle as water is not generally seen as a problem, certainly not in relation to scarcity. This perception constitutes a barrier to professionals as well as the public initiating sustainable water management measures. Respondents cited how the case study projects had opened their eyes to the problems as well as solutions about water management.
4	**Better incorporation of water management into development planning** • planning policy should be developed to ensure that local issues are fully understood with respect to water and drainage infrastructure before planning permission is given for new development. The water cycle study guidance report (EA, 2008) is helpful in this regard • in particular, consideration should be given to water resources and drainage in regional plans and local development frameworks through explicit involvement of both the EA and water companies • attention should also be given to the cumulative effect of new housing or infill development if beyond the existing planning framework • inclusion of water-related professionals in development planning processes should help planners recognise and address the sustainability issues of specific developments.
	Basis of recommendation: water management innovations are not common in the UK. Where they do occur, such as in the case studies, they are usually delivered due to the tenacity of keen individuals. This applies even in areas where there are clear water management issues, such as water scarcity.
5	**More case studies of SWCM** • more case studies should be developed to promote and provide examples of SWCM • there should be greater encouragement for developments to become showcase sites on how SWCM is achieved.
	Basis of recommendation: there are few examples of SWCM in the UK. Case studies can be used as lessons on what practices are most successful and provide recommendations for good practice. They can also demonstrate difficulties experienced in the development to help stakeholders overcome and learn from them.

5.7 CONCLUSIONS

- taking up SWCM is less likely if it is a voluntary process
- local authorities are well suited to play a vital role in adopting SWCM. The role of local authorities in adopting SUDS is crucial and the Flood and Water Management Bill 2009 proposals are welcomed
- if reputable organisations are involved or champion SWCM, people will have greater trust and levels of participation.
- adopting SWCM is easier where existing stakeholder relationships are already in place
- community-based SWCM schemes can work provided concerned stakeholders are fully engaged and committed
- there is a huge education requirement with respect to the water cycle, water scarcity, water value and using water wisely

- modest financial support can boost learning and innovation in relation to sustainable water management
- changing behaviour through persuasion involves effective communication, which requires a two-way exchange of understanding. Changes in public behaviour are only likely to occur if people have first felt their needs, interests and priorities have been recognised and taken on board in the corresponding industry's actions and messages
- there is still considerable variability between water companies and how they achieve their mandatory duty to promote the efficient use of water
- SWCM communications should not be expected to deliver change overnight. The public need to hear messages on a repeated basis and through different media for those messages to make a difference
- evaluating the achievement of interim goals (ie primarily the achievement (or not) of changes in perceptions about water) provides a basis for checking whether and how specific campaigns are working, and for informing the development of better campaigns.

5.8 FURTHER RESEARCH AND GUIDANCE

- **further case studies on the adoption of SWCM in developments would be helpful**
- **greater inclusion of the assessment of stakeholder perceptions and behaviours before and after SWCM initiatives would be useful**
- **specific guidance and training would greatly benefit those involved in stakeholder engagement.**

6 Tools for decision making

This chapter:

- presents the tools that were developed during the WaND research project to manage the design and planning of urban stormwater management components and monitoring performance of sustainable drainage systems
- discusses how the tools help facilitate sustainable water management, explains their functionality and how they can be developed.

6.1 INTRODUCTION

This chapter sets out the tools that were developed during the WaND research project. Most were initially developed to support the researchers through the decision making process, and to compare the different technologies. At the time of writing, they exist in varying states of development and usability by practitioners. The tools that have not been specifically developed for practitioners have still been included as they could prove to be very useful, with potential to be adopted and further developed by commercial organisations.

This chapter summarises the main decision-support tools developed within the WaND research project, specifically for water cycle management in new developments. Table 6.1 identifies the purpose and appropriate application stage for each tool, the potential end user, the development status of each tool and the contact organisation that can provide access to and support for the particular tool. More specific contact details for these organisations are given in Appendix A7.

The tools focus on providing the necessary information for implementing sustainable technologies once a site has been chosen rather than the more strategic elements of planning. However the site screening tool (SST), the demand forecasting tool and to some extent the project assessment tool (PAT) are aimed at the more strategic stage of site selection. They have been included to demonstrate how they can help achieve sustainable water cycle management (SWCM) at the strategic level.

6.2 THE WaND DECISION SUPPORT TOOLS

These tools were based on detailed reviews of current and innovative water management practices and technologies that have been carried out under the WaND project. These identify the main design and performance criteria/indicators and assess the suitability of these techniques for new developments and their potential contribution to sustainability. Research also identified the social, economic and health aspects and potential constraints of these technologies. The more quantitative aspects of these reviews and other sources have been collected into an excel-based technology library included in the WaND portal. This excel-based technology library was developed for the urban water optioneering tool (UWOT), (Section 6.1.3). A short description of the tools and their capabilities is included in the following sections.

Table 6.1 *WaND decision support tools*

Name of tool	Purpose	Application stage	Intended users	Tool development status	Contact
Project assessment tool	Analyse the sustainability of a proposed project at an initial/high level.	Strategic but should be consulted throughout the development process	Developers and their consultants as well as planners and local authorities	Operational prototype (available on the WaND portal on CD-Rom with this publication)	University of Sheffield
Site screening tool	Identify areas suitable for new developments	Strategic planners and local authorities	Operational prototype	Not yet developed as an end user product	University of Exeter
Urban water optioneering tool	Identify combinations of water technologies improve the performance of a specific new development	Master planning	Developers and their consultants	Technology library and lite version available with this guide Full prototype available on request from University of Exeter	University of Exeter
Suitability evaluation tool	Identify where to locate new water technologies within a development site	Master planning	The researchers and potentially developers and their consultants	Research prototype Not yet developed as an end user product	University of Exeter
Demand forecasting tools	Forecast water demands under a range of scenarios	Strategic site selection	Water resource planners	Operational prototype available on the web	University of Leeds
Stormwater management tools	Identify appropriate stormwater management methods and evaluate their performance	Initial scoping stage	Developers and their consultants	Operational Available from <www.uksuds.com>	HR Wallingford
Greywater recycling tools	Size and evaluate the performance of greywater recycling technologies	Master planning	Researchers of the WaND project	Research prototype Not yet developed as an end user product	University of Exeter

6.2.1 Project assessment tool (PAT)

The project assessment tool (PAT) can be used to evaluate the proposal for a new development and encourages multiple stakeholders to discuss and articulate their preferences and concerns and set out sustainability criteria. The criteria indicators framework within the tool is intended to encourage the wider consideration of sustainability issues at the preliminary stage of a development proposal. The tool should be used at the earliest stage possible in the planning process, but can also be used to evaluate the sustainability of a site at different stages through the development.

PAT was based on two pre-existing assessment frameworks: SWARD (Ashley *et al*, 2004) and SPeAR® (developed by Arup) as discussed in Chapter 3. Within the SWARD framework, the principles of sustainability are simplified into constituent categories of criteria that should be equitably fulfilled to provide a balanced decision. The criteria are then sub-divided into primary and secondary criteria that are specific to the decision at hand (Table 6.2).

Table 6.2 *Categories and primary criteria of sustainability evaluation used by PAT*

Category	Primary criteria
Economic	• life cycle costs, the whole-life costs of investment, resource extraction, production, construction, end use and decommissioning • willingness to pay for an attribute incorporated at design stage • affordability to customer • financial risk exposure, the risk of loss associated with investment.
Environmental	• resource use, the land, energy, chemical, material and water resource use during the whole-life of investment • service provision: water consumption, leakage and reuse • environmental effect on land, water, air and biodiversity.
Social	• effect on risks to human health • acceptability to stakeholders of the investment scheme • participation and responsibility: stakeholders participation and responsibility in sustainable behaviour • public awareness and understanding: awareness of sustainable development and implications of behaviour • social inclusion: by water utility actions.
Technical	• performance of the system, quality of effluent treated in relation to required standards • reliability of the system in performing its functions • durability, the level of accommodation in design of the system • flexibility and adaptability, the ability to add or remove from the system.

PAT is a simple, subjective assessment adopted within a spreadsheet model. The opinions can be recorded within the PAT spreadsheet to provide an audit of the whole decision making process. This requires written explanations of the scores given to each criterion. Scores are allocated between +3 and -3 with 0 indicating status quo.

For incorporation of more detailed information, PAT can use data generated by higher level and more detailed decision-support frameworks and software. Aggregated bar charts provide visual representation of the evaluation of both primary criteria and sustainability categories. The project assessment tool is described in more detail in Hurley *et al* (2007b).

The PAT needs no further development. It is available with this project and requires the stakeholders to develop their own set of specific criteria.

6.2.2 Site screening tool (SST)

The site screening tool (SST) uses the categories of primary criteria (economic, environmental, social and technical) to establish the best location for new urban developments. It is a GIS-based, decision-support tool that can prioritise sites suitable for new urban areas and to screen out unsuitable sites, using a multi-criteria approach (Butler *et al*, 2005).

For this tool the user identifies specific objectives by the selection of sustainability criteria (a set of example criteria have been included in Tables 6.3 to 6.5). These objectives are required because they are applied to a specific geographical area. This area should be large enough to identify the best location available and it is suggested that it is applied at a regional scale.

Table 6.3 **Environmental sustainability criteria and indicators**

	Criteria	Indicators (Spatial Attributes)
General	Effect on sites of ecological and environmental significance (PPG 9)	Location of sites characterised as SSSI, SAC, SPA, RSAR, NNR, sensitive waterway or ancient woodland Distance from sites of ecological and environmental significance
	Influence of new development on established land-use	Location of existing land-use sites
	Effect on countryside and urban regeneration (PPS 3)	Location of brownfield sites Location of greenfield sites
Groundwater	Groundwater quality and recharge and abstraction areas Development near vulnerable aquifers Water Framework Directive (2000/60/EEC)	Distance from source protection zones Location of areas with sustainable abstraction status
Surface water	Effect on nutrient-sensitive areas Urban Waste Water Treatment Directive (1/271/EEC)	Location of nutrient-sensitive areas
	Effect on natural stormwater drainage patterns (stormwater management and design manual)	Distance from streams

Table 6.4 **Social sustainability criteria and indicators**

Criteria	Indicators (spatial attributes)
Gap between the poorest communities and the rest (PPS3)	Location of most deprived communities in the UK according to the index of total deprivation
Social isolation and socio-economic inactivity (PPS 13)	Distance from urban centres and trade areas

Table 6.5 ***Economic sustainability criteria and indicators***

Criteria	Indicators (spatial attributes)
Cost of site development Ease of transportation (PPS 13, PPS 3)	Ground slope Proximity to transportation nodes and branches
Access to safe drinking water	Proximity to reservoirs, abstraction wells and water supply networks
Flood risk. Selected sites need to be included in flood zones 1 and 2. (PPS 25)	Distance to floodplains

The SST system architecture integrates two widely used software platforms ArcView GIS and Matlab into a user-friendly interface of a spreadsheet. The user requires GIS and Matlab software to use the tool since, even though the inputs are in an excel spreadsheet, the outcome is a map identifying the preferred sites of new urban developments.

6.2.3 Urban water optioneering tool (UWOT)

The urban water optioneering tool (UWOT) manages the comparison of different water management technologies (including water saving, recycling, treatment and drainage) at different scales. It addresses the difficulty of selecting the most suitable technologies with the focus on the numerous technologies that are all separate entities, but need to be considered cumulatively. To help this, the WaND project developed UWOT, which provides a range of technology combinations. It then ranks these combinations depending on user-based criteria. This allows the user to make an informed decision on the water-saving strategies that are most beneficial for their development.

A knowledge base was developed to collect available water management technological options. This knowledge base, termed the technology library, was adopted in multiple spreadsheets, one for each category of technological option (baths, showers, greywater treatment systems etc) and contains data and information on their major characteristics and performance. The information included in the library is based on the environmental, economic, social and technical indicators introduced in earlier sections. Some of these indicators are quantitative (eg installation costs for a particular tap) while others are qualitative and require input in the form of expert judgement (eg social acceptability of a particular greywater recycling scheme).

Box 6.1 *Using the site screening tool in the Humber sub-region*

The site screening tool was applied to the Humber sub-region in the UK. This sub-region covers an area of 3517 km^2. Its population was estimated to be 870 600 (as at 2002). It is geographically and socio-economically diverse.

The Humber estuary is the sub-region's characteristic physical feature and potential source of economic development. It provides a major trading outlet towards Europe and the adjoining UK regions. The city of Hull acts as the sub-regional capital and is the primary economic, civic and cultural centre. To explore the functionality of the decision-support system developed and its capability to support users with different objectives and prioritisation, a scenario approach was adopted. The scenario is an example of using SST to support the aims of regional planning guidance. A review of the UK Regional Planning Guidance (2000) and the UK Regional Sustainable Development Framework (2003) for Yorkshire and the Humber indicates several attributes within the case study area that can be related to some of their aims and objectives. These formed the basis of the multi-criteria analysis. The planning guidelines suggest the following order for attributes:

1 Distance from transportation networks.

2 Groundwater availability.

3 Distance from sites of environmental significance.

4 Distance from trade activity zones.

5 Distance from derelict and brownfield sites.

6 Social inclusion.

The result of this combination of criteria is the suitability map that is displayed in Figure 6.1. The suitability ranges indicate the influence of the high ranking criteria such as groundwater availability. The area located on the aquifer section where groundwater is still available receives the largest suitability values (the dark area). The criterion that demands proximity to transportation networks also plays an important part in the final result, especially in the more remote parts to the north of the study area. The areas situated to the south of the Humber estuary are on average more suitable in comparison to the north bank. This is because the south part of the region being located in the Humber trade zone, while the north part features many areas of environmental significance.

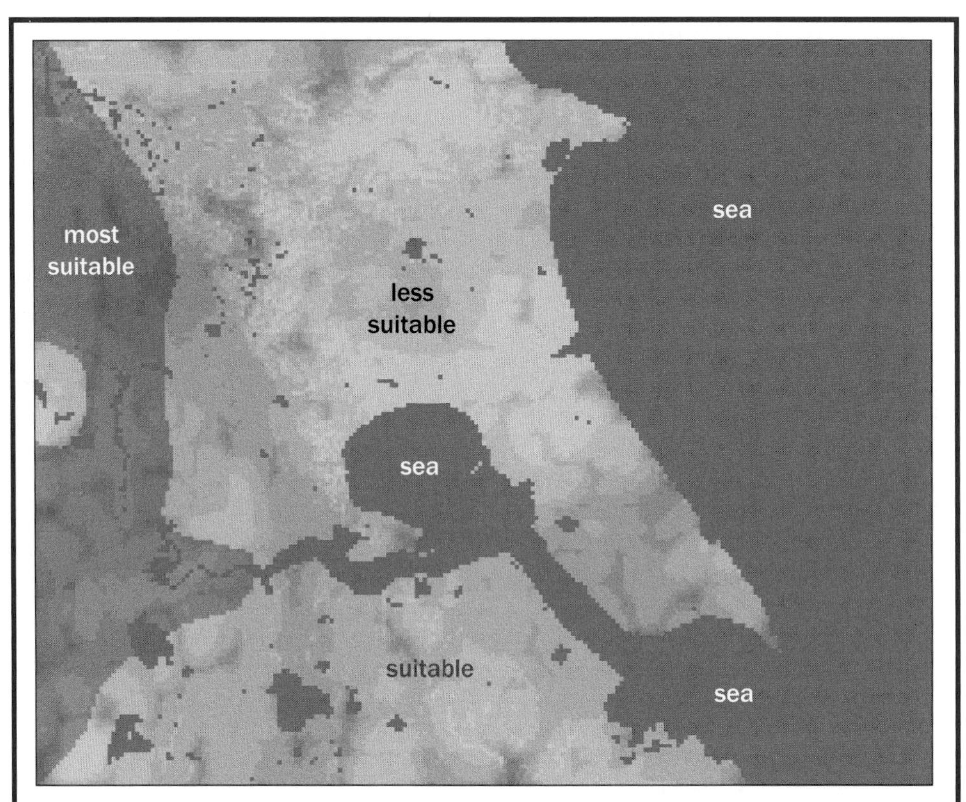

Figure 6.1 *Map of Humber estuary showing suitability areas (this image is also explained in more detail on the WaND portal, see Appendix A1)*

A different suitability map would have been created, for different combinations of criteria, indicating different agenda from stakeholders. Two further scenarios (for an equal importance for environmental criteria and an eco-centric approach) have been investigated in Butler *et al* (2005). In all, for the Humber sub-region, the possibility of environmental effects greatly reduced the overall suitability, even though the study area was socio-economically advantageous.

Complementing the sustainability indicators, the technology library also contains other parameters, on technical and operational characteristics for each technology required for the calculation of the water balance for the total modelled urban water cycle (eg water use per flush for a specific type of toilet). The technology library can be updated with new technologies and their performance characteristics without any need of updating the tool.

UWOT integrates Simulink/MATLAB and Microsoft Excel into a powerful user-friendly decision-support tool. Excel is used for data input, storage of technology options characteristics and graphical outputs. MATLAB/Simulink computes the water mass-balance model, for the combination of technological options selected for evaluation by the user and processes its outputs in a form suitable for sustainability assessment, optimisation and visualisation in excel.

UWOT can be run in two modes:

1 Assessment: in this mode, UWOT evaluates a user-defined configuration of an urban water management system (for a specific urban development).

2 Optimisation: the system suggests a configuration, subject to the boundary conditions of the problem (including size of development, occupancy etc).

Each configuration is evaluated in view of the sustainability criteria discussed previously, and the results of the evaluation are communicated to the end user via a spider plot (see Figure 6.2). The tool has been successfully demonstrated on the Elvetham Heath case study site (see Box 6.2).

Box 6.2 ***UWOT application in Elvetham Heath***

Elvetham Heath is a 126 ha development situated in Hampshire in the south-east of England. It contains a natural reserve, car parks and 62 ha of a residential area planned to accommodate 6000 residents.

UWOT was applied to the problem of selecting water-related technologies at Elvetham Heath (EH), to test the water-balance model, technology library and functionality of the decision-support interface. UWOT was first applied to simulate the present situation in EH and afterwards to assess the sustainability of various alternative water management practices that could be applied in the new households during the development's planned expansion.

Some UWM practices have been successfully adopted in Elvetham Heath in the form of SUDS. The village runoff drainage system includes swales/retention ponds that help manage flood risk and also contribute to the aesthetic value of the village. Scenarios were developed to provide a basis for considering alternative development possibilities in the case study area. Only one of the scenarios developed (water save) is presented against a benchmark scenario.

The benchmark scenario, or "business as usual", resembles residential developments using traditional water-intensive technologies. The scenario uses population and household data specific to Elvetham Heath and makes use of conventional technologies at the household level assuming no rainwater harvesting, greywater recycling or SUDS. It also assumes that all water services, ie water supply, stormwater and wastewater conveyance and treatment, are delivered using centralised systems. Because rainwater harvesting and greywater recycling is not allowed in this scenario, the only source of water supply is potable water from the water service provider (through external water mains).

The mean per capita consumption corresponding to the development under the benchmark scenario, calculated over a simulated period of 10 days, is about 168 litres per head per day (l/h/d). This is about 10 per cent higher than the 150 l/h/d, which is identified as a typical value for the UK (see also Butler and Makropoulos, 2006). This is attributed to the case study having a significant amount of gardens and associated outdoors water use. Wastewater generation is about 85 per cent of the water demand, which is in agreement with generally accepted losses of five per cent within the house (Butler and Davies, 2004) augmented by the 10 per cent of demand due to gardening (water used in garden irrigation is not contributing to the measured wastewater flow). The results provide confidence that the model is well constructed and able to correctly simulate the water cycle within a development.

The water save scenario is based on the same baseline data (number of houses, occupancy, pervious and impervious areas), but uses water-saving technologies at the household level, for example, dual-flow WC cisterns and low-flow showerheads, which are selected (manually) by the user. This scenario is realistic as the technologies employed are already developed and both the market and society are progressively adopting them. Similar to the benchmark scenario, the water save scenario assumes no centralised greywater recycling and rainwater harvesting and the only source of water is the water service provider external supply. The water save scenario shows a decrease in demand for potable water, indicating a per capita consumption of 109 l/h/d (level 2 of the Code for Sustainable Homes), which is in agreement with what literature suggests as possible following the installation of water-saving devices. Figure 6.2 presents a sustainability assessment of the water save scenario, comparing it to the benchmark scenario.

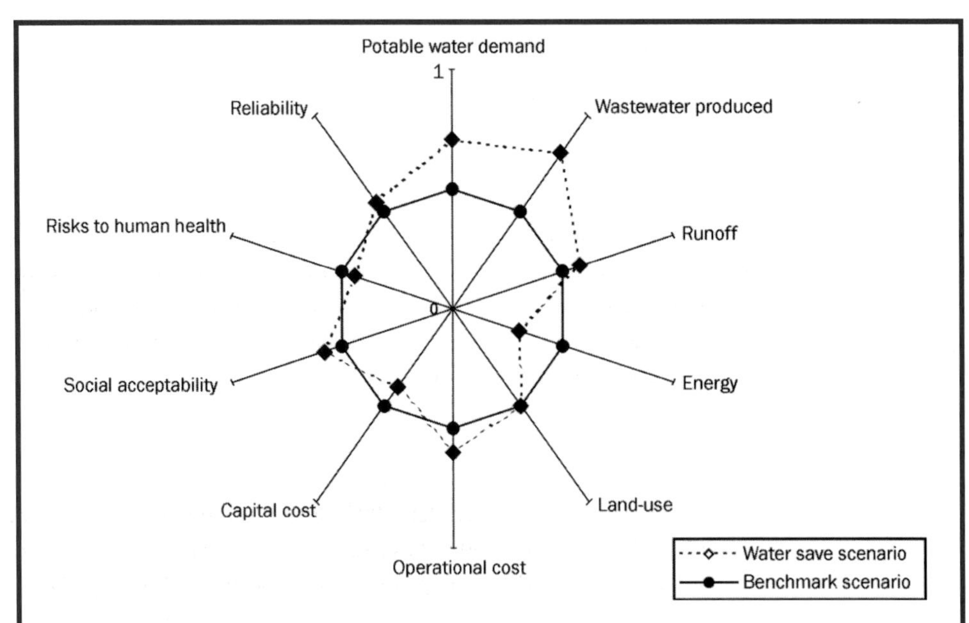

Figure 6.2 *A comparison of the water save scenario with the benchmark scenario against multiple sustainability objectives.*

A scenario value lower than the benchmark indicates poorer performance and lower sustainability. Similarly, high sustainability values are placed outside the benchmark values. In this example, it can be observed that indicators "potable water demand" and "wastewater produced" have improved at the expense of "energy". Although the actual values of these indicators would depend on the specific technologies used, it is suggested that changes in the water cycle, through the introduction of water managing technologies have implications on the entire system and can affect a series of (sustainability) metrics.

It is argued that optioneering tools have great potential to become part of urban water management planning toolkits as there is a move toward decentralised, integrated, context-specific solutions to address issues of sustainability.

Further details on UWOT can be found in Makropoulos *et al* (2008) and Sakellari *et al* (2005). This tool is operational and is available on request from the University of Exeter.

6.2.4 Suitability evaluation tool

The aim of this tool is to assist in the optimal siting of water management strategies (eg SUDS) in new developments. The main output is a suitability map, which suggests locations for water technologies on a proposed site. The tool uses a spreadsheet based interface where the user identifies multi-criteria decision analysis. The user can select a particular water management strategy (eg SUDS), identify the constraints to taking up this water management strategy and define the cut-off values that will indicate whether the application of the strategy is acceptable or not. Decision makers identify their preferences from the following inputs:

- selection of a strategy
- selection of attributes
- choice of weighting method
- weight assignment
- choice of aggregation method
- attitude towards risk
- area requirements.

The information provided is then processed in MATLAB, which creates a map of the results to highlight where the technologies should be located on site.

This tool needs further development because it was developed for the WaND research and not for the end user.

6.2.5 Demand forecasting tools

The demand forecasting tool is a strategic water resource tool. It is aimed at water resource planners to provide them with forecasts of future water needs for large-scale areas. The application of the tool allows stakeholders to investigate and quantify the effect of a new development in relation to available water resources and water-saving technologies on overall water demand.

Factors influencing future domestic water demand include population growth, household size, lifestyle, climate change, affluence and taking up new water-using and water-saving technologies (Memon and Butler, 2006). Several water resources planning documents, including Samuels *et al* (2006), have identified uncertainties in demand forecasting as one of the main issues in developing cost-effective security of supply plans. Developing reliable water demand forecasts is a complex task, because it requires taking into account the dynamic variation in population migration trends, socio-economic profiling, climate change and the roll-out rate of innovations in water-saving micro-components. Reliable forecasts are important for planning and decision making processes.

To help this process the WaND research project developed three major research tools each serving a different purpose (see Table 6.7). The microwater and macrowater tools have been combined in one spreadsheet for water resource planners. The spreadsheet allows the user to input data and choose parameters depending on scale, site location and other factors. The outputs are provided in the form of tables and graphs.

Table 6.7 *Demand forecasting tools developed in WaND*

Tool	Aspect	Details
MicroWater	Description	Microwater is a scenario based, bottom-up model based on current and future micro-component values, designed to reflect changes in technology (volume) and social norms (ownership, frequency). It also uses government targets for house building and water efficiency and Ofwat targets for metering.
	Timescale	2001 and 2031 (two points of comparison)
	Spatial scale	From street to national level
	Application	The tool has been applied for the four government office regions
	Simulation method	Micro-components and scenario modelling
MACROWater	Description	Macrowater is a scenario-based, top-down model based on government targets for house building and water efficiency and Ofwat targets for metering. New and existing housing stocks are processed separately as they have different water-efficiency profiles. The effect of climate change is taken into account. The tool uses demand patterns from domestic consumption monitors (DCM) from the water companies calibrated to match 2001 Ofwat statistics, combined with housing, household and population projections.
	Timescale	2001–2031 in five year intervals
	Spatial scale	All local authorities in a development zone/water company/resource zone
	Application	Thames Gateway study area, four government office regions
	Simulation method	Additive accounting, iterative proportional fitting, fuzzy matching (for missing company data) and scenario based projection
MicrosimWater	Description	Microsimwater is a scenario based micro-simulation model that explores the water demand dynamics affected by government planning, private investment, climate change, behaviour and socio-economic change. Household and population data from UK Census 2001 and water consumption survey data are fused to provide a baseline micro database. This micro database is used to produce base level water demand estimates for small areas (middle level super output areas (MSOAs)) in a development zone
	Timescale	2001–2031 in single years
	Spatial scale	All MSOAs in a development zone. A MSOA is about the size of a ward
	Application	Thames Gateway study area
	Simulation method	Iterative proportional fitting, combinatorial optimisation, statistical matching, micro-component, demographic micro dynamic and scenario based projection

Further details on their architecture and application are provided on the WaND portal.

The combination of the microwater and macrowater into one tool is available to use from the University of Leeds water website <www.water.leeds.ac.uk/wand>. This research tool provides evidence for accommodating future water demands.

6.2.6 Stormwater management tools

The stormwater management tools were developed for developers and planners. The tools are intended for the initial assessment of water on site to quickly establish what storage volumes are required. The cost of drainage depends on storage volume and these tools quickly estimate the volumes that will give a developer a greater understanding of the cost implications. The main aims of the tools are establishing:

- storage requirements for drainage systems
- the drainage issues and SUDS components that are likely to be most appropriate.

This approach resulted in the development of several tools that can be broadly classified into two groups:

1 Tools and guidance for preliminary drainage design.
2 Tools and guidance for the evaluation of sustainability of detailed drainage design.

Tools and guidance for preliminary drainage design: two tools have been produced:

1 **Site suitability evaluation tool for SUDS implementation:** uses site-specific characteristics of the development to provide guidance on the applicable use of SUDS. The interface allows the user to answer eight simple questions about the site. The results highlight the SUDS methods that are most appropriate and a brief description of the SUDS.

  This is a very useful tool especially for the developer when deciding the SUDS methods that are most appropriate for the site.

2 **Surface water storage requirement assessment tool:** estimates the storage volume requirements that are needed to meet normal good practice criteria in line with Environment Agency and CIRIA guidance on sustainable drainage. The tool consists of a map that enables the tool user to locate the site of the proposed new development and input development site-specific information including: proposed percentage of impermeable area, rainfall data, soil type, climate change factors, percentage of impervious surfaces and desired treatment scale. Volumetric requirements are calculated in terms of long-term storage, attenuation storage, treatment storage and interception storage.

  This tool is useful for understanding how much water will be on site and the volume of water that could be stored for GWS or RWH.

Tools and guidance for the evaluation of sustainability of detailed drainage design. Three tools have been developed:

1 **Joint probability assessment tool:** the aim of this tool is to address the problems associated with a large rainfall event and the dependency between urban runoff and the catchment it is located within for conveyance and discharge. This is particularly relevant where a SUDS system has a pond or basin located close to a floodplain or river (where the change in water level is below 5 m).

  During an event the critical duration for the drainage system is similar to that of the catchment drainage (between six and 24 hours or more). When extreme events take place the potential high river levels will affect the drainage of a site. The dependency is likely to be high and if the discharge from the site is constrained by high river levels, then careful consideration of the hydraulic performance of drainage systems for extreme events needs to be made. This tool can provide some guidance on return periods where dependency between events should be taken into account. It can also provide initial guidance on joint dependency for a site and can indicate if a more detailed studied of records and drainage systems is needed.

2 **Infiltration design tool:** Part H of the Building Regulations advises the designer to first consider infiltration as a means of stormwater disposal before the use of piped drainage. Infiltration should not be used where the ground is relatively impervious, or has high groundwater table or where there is a risk of contaminating groundwater.

  This tool uses a simple excel input sheet to calculate the sizing of infiltration soakaways and trenches and can design:

  1 Rectangular soakaway.

2 Ring soakaway.

3 Infiltration trench.

Environmental assessment tools for a drainage system: these may be regarded as sustainability evaluation tools. The WaND research project has developed two tools:

1 **Water quality assessment tool:** this is a qualitative approach to assessing stormwater runoff from a site once it has been processed by the proposed SUDS systems. It assesses the adequacy of the proposed SUDS system in providing treatment to the stormwater runoff. The tool takes into consideration the relative performance of SUDS, type of land-use area and sensitivity of receiving waters.

   This tool is not based on comparing the runoff water quality with greenfield conditions, nor does it attempt to compute pollutant concentrations. It assumes that runoff is only contaminated from paved surfaces. Runoff to infiltration is excluded from this calculation.

2 **Hydraulic performance assessment tool:** this tool compares the drainage system's response to rainfall events before and after the development on site. The objective is that runoff rates post development should not exceed those from a greenfield site. This assessment is applied to varying storm events (a time series of extreme and frequent events) and not to a single design storm. Performance is measured against the following parameters:

   - peak flow rates
   - runoff volumes
   - volume of runoff that infiltrates.

The stormwater management tools have simple parameters to input. The excel interface is easy to navigate and gives quick results. They provide simple results for the initial design stages of a development and guidance on how the site should focus on managing stormwater. These tools are operational and can be accessed on the UKSUDS website <http://www.uksuds.com/>.

6.2.7 Greywater recycling tools

Greywater recycling has been used internationally in water-stressed areas, but so far has not found widespread use in the UK. In principle, it helps in the preservation of high quality fresh water supplies and potentially reduces the pollutant load to the environment. Greywater is defined as water that is slightly contaminated by human activities (eg water from personal washing, showers, baths and possibly washing machines), which may possibly be reused after suitable treatment. Treated greywater, in a household context, is used for toilet flushing and/or garden watering (Liu *et al*, 2007). A typical system consists of three components:

- a tank to store raw greywater
- a pump/treatment unit
- an overhead tank to store treated greywater (green water) before supply to the toilet.

Practical experience of single-household systems has been mixed, and it is argued this is partly due to poor system design. The WaND project developed several tools to investigate aspects related to the environmental, economic and design performance of greywater systems to increase confidence in system performance and specify the appropriate scale of provision. These tools were developed as part of the WaND research and were not intended as tools for the end user, but they provide important results to help the decision making process for SWCM. These tools and their results are:

1 **Storage tank sizing tool:** this model represented the main elements of any recycling system (inputs, outputs, tank sizes and treatment capacity) to allow determination of storage tanks volumes to be calculated in a more rational and realistic way. It provided insights into the relationship between performance and system configuration. The model was used to assess system performance from both a volumetric and water quality point of view.

   The main model was developed using an object-based approach adopted in MATLAB. The model is linked with a purpose-built user interface in excel via an excelink module. Further details on the model's architecture and tool use are available in Liu *et al* (2007).

   Results from the volumetric analysis indicate that a higher water-saving efficiency can be achieved for a system with larger green water tank sizes, but that greywater tank volume is relatively insignificant above a certain minimum. The quality-based analysis highlights that although larger volume tanks produce higher water-saving efficiencies, smaller volume tanks (offering short hydraulic-retention time) are needed to secure good water quality.

2 **The life cycle analysis tool (LCA):** this tool investigated the environmental effect of greywater treatment technologies in the WaND research. Two conventional (reed beds and membrane bioreactor (MBR)) and two innovative (GROW and membrane chemical reactor (MCR)) technologies have been considered. LCA outputs were generated for each technology for a range of development scales (from 50 to 2000 households) using conventional LCA protocols. The LCA outputs for each technology were processed using an adaptive neuro-fuzzy inference system in MATLAB to develop a set of technology-specific generic tools. These tools have the ability to capture the influence of development scale, energy use and material consumption (in technology construction and operation) on vital environmental effect indicators (eg abiotic depletion, global warming, human toxicity, acidification and eutrophication). The results indicate that per capita environmental effect is inversely proportional to the development scale and energy-intensive technologies have high environmental effect. Further details can be found in Memon *et al* (2007).

Figure 6.3 shows a graphical representation of the contribution of stainless steel and PVC (used in the membrane-based treatment technologies) towards environmental impact indicators (global warming and human toxicity).

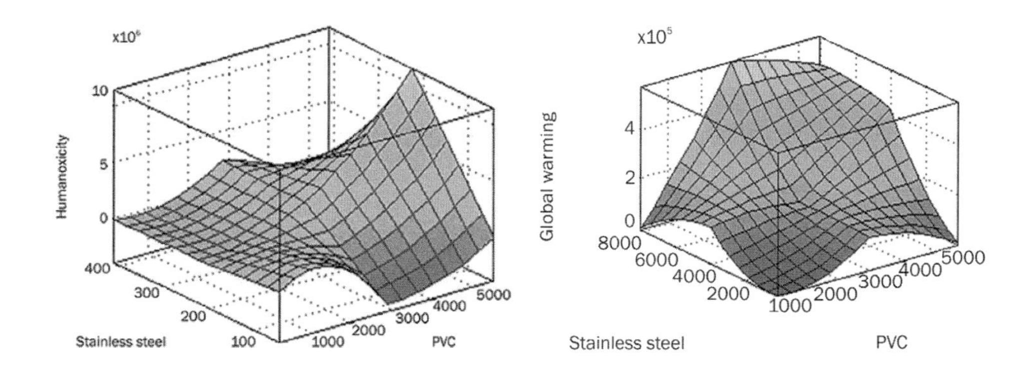

Figure 6.3 *Example output from LCA tool showing environment effect of materials used in membrane based technologies*

3 **Economic assessment tool:** this tool was aimed at quantifying the whole-life cost and performing cost-benefit analysis of greywater recycling systems using the development scale, technology attributes and users' water consumption patterns as the main influencing variables. The tool was used to investigate the influence of:

- system design life
- water-saving appliances
- water pricing strategies
- system maintenance regime
- climate change
- system efficiency.

The tool consists of four integrated modules:

1. **Input module:** the input module provides a mechanism to select parameter values necessary for defining the scale of greywater recycling system (ie single-household, medium or large scale) and efficiency of the system. The scale of system is assumed to be dependent on the average number of residents and days when greywater is produced and recycled.
2. **Water flow module:** the simulation results from this module provide input for cost calculations and quantify water-saving potential. The module has two components: greywater generation and consumption. The greywater generation and consumption quantification has been carried out using micro-components associated attributes (ownership and frequency of use for each appliance and nominal water consumed per use by each appliance contributing to greywater).
3. **Cost quantification module:** this module calculates the net cost of a greywater recycling system by taking into account the capital cost, regular and unplanned maintenance and operation costs and savings resulting from greywater reuse.
4. **WLC assessment module:** this module computes the whole-life cost (WLC) of a greywater recycling system as a function of total capital cost and the net present (NPV) of running (O & M) cost and decommissioning cost incurred at the end of the design life of the recycling system.

The tool was applied using cost data from two case studies. The results indicate that the large scale greywater recycling systems are financially viable. Also, low-flush toilets have a negative effect on cost-benefit ratio, because they reduce water use potential of treated greywater. Further details on the tool development strategy and its use for quantifying the whole-life cost for two case studies can be found in Memon *et al* (2005).

6.2.8 Health impact assessment tool

SWCM technologies ideally should have no health implications or any effects on quality of service. The WaND research project has developed a generic methodology to perform health impact assessment of innovative water management strategies and technologies. The methodology employs the health impact assessment (HIA) framework as described in WHO (2006). The methodology is aimed at assessing and quantifying possible effects (positive and negative) of water management options as a function of disability adjusted life years (DALYs). DALY is a health-gap measure that extends the concept of potential years of life lost due to premature death to include equivalent years of "healthy" life lost by virtue of being in states of poor health or disability. The DALY combines in one measure the time lived with disability and the time lost due to premature mortality.

The HIA methodology developed in WaND is based on a series of hazard checklists developed for a range of water management options. Further details on the methodology development and its application are given in the WaND portal.

The methodology was applied to a range of water management options including rainwater harvesting, greywater recycling and SUDS. The results were then compared with the health effect of a scenario based on a new development experiencing a conservative and non-conservative drought-like situation every five years. A summary of results is shown in Figure 6.4. The results indicate that the health-related consequences for most of the properly managed water management options are within limits set by WHO.

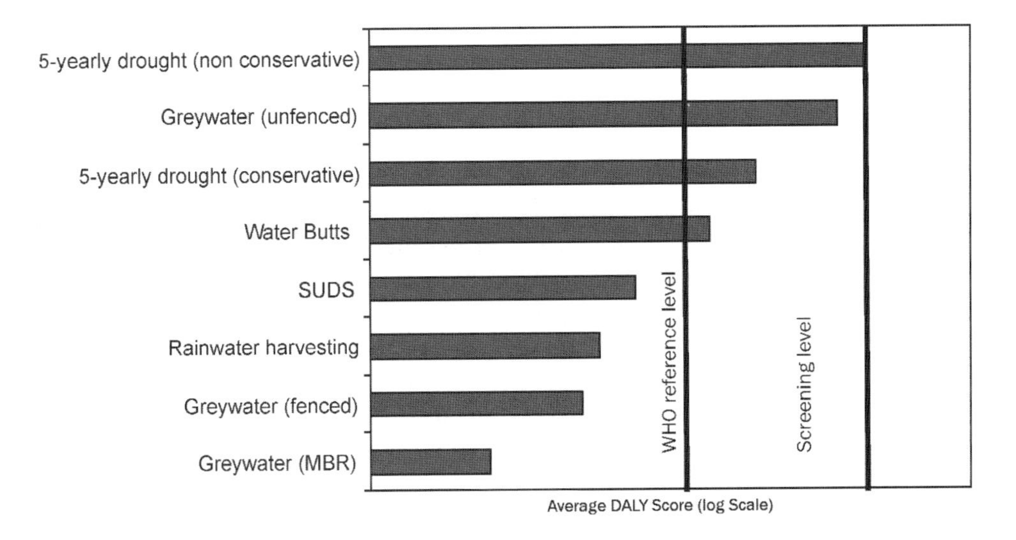

Figure 6.4 *Mean DALY scores for each of the selected water management options (on a log scale) in comparison with the WHO and screening reference levels*

As part of the health impact assessment procedure development, risk management advice for a range of water management options was also developed and can be found via the WaND portal.

6.3 USE AND DEVELOPMENT OF THE TOOLS

The tools described have the potential to be developed to create a toolkit for achieving SWCM. The WaND project has identified the various areas where tools are needed, what the tools can deliver and development of them has begun. However, until a user-friendly interface is developed for the tools, they are unlikely to be taken up by practitioners. The tools that show potential for being developed are the SST, SET and UWOT.

These tools can be used singly or in combination, to support assessment across all the stages of a water cycle study. The process can be seen in Figure 6.5 with a supporting scenario to explain the process:

1 The first reference for all stakeholders is the WaND portal. The decision maker uses the portal to learn about the decision process, navigate through the necessary stages and tasks and identify tools and approaches relevant to each stage.

2 The decision maker may then undertake a decision mapping exercise to better understand the process, stakeholders and information flows that characterise the problem and include them in the process.

3 The project assessment tool is used to evaluate the proposal for a new development

and allow multiple stakeholders to discuss and articulate their preferences and concerns. The criteria indicators framework created at this very preliminary stage of the proposed new development will be propagated and modified when necessary.

4 Following a decision to build a new urban development the screening tool is used to support the selection of a particular area, as the "best" for a new development, within a particular region of interest. The tool is able to incorporate the relevant technical and environmental issues already gauged in the project assessment stage, and use them for selection while also taking into account broader requirements related to the economic and social opportunities and constraints of the region. It prioritises areas suitable for new urban areas and screens out unsuitable sites using multi-criteria analysis informed by the decision maker's preferences and constraints.

5 Once the site is defined, the optioneering tool (UWOT) is used to explore the potential application of technologies, included in a technology library, to the design of the total urban water cycle. This process can take two forms: (a) the user can suggest appropriate technology setups and UWOT will evaluate the performance of their ensemble for a range of sustainability indicators, or (b) UWOT can be run in optimisation mode to suggest an appropriate mix of technologies (from taps to recycling and treatment systems) that are compatible to each other and perform better for a set of pre-defined sustainability objectives (cost, environmental effect etc).

6 UWOT can be fed with information from the use of other flow-specific tools (eg water demand predictions from the macro and microwater tools or a pre-selection of possible drainage solutions from the SUDS tools). It can encourage the application of downstream tools, such as the GIS-based suitability evaluation tools, which can assist with locating each of the technologies of choice within the new development. The final solution can be used as the starting point for more detailed simulation using, for example, commercial tools.

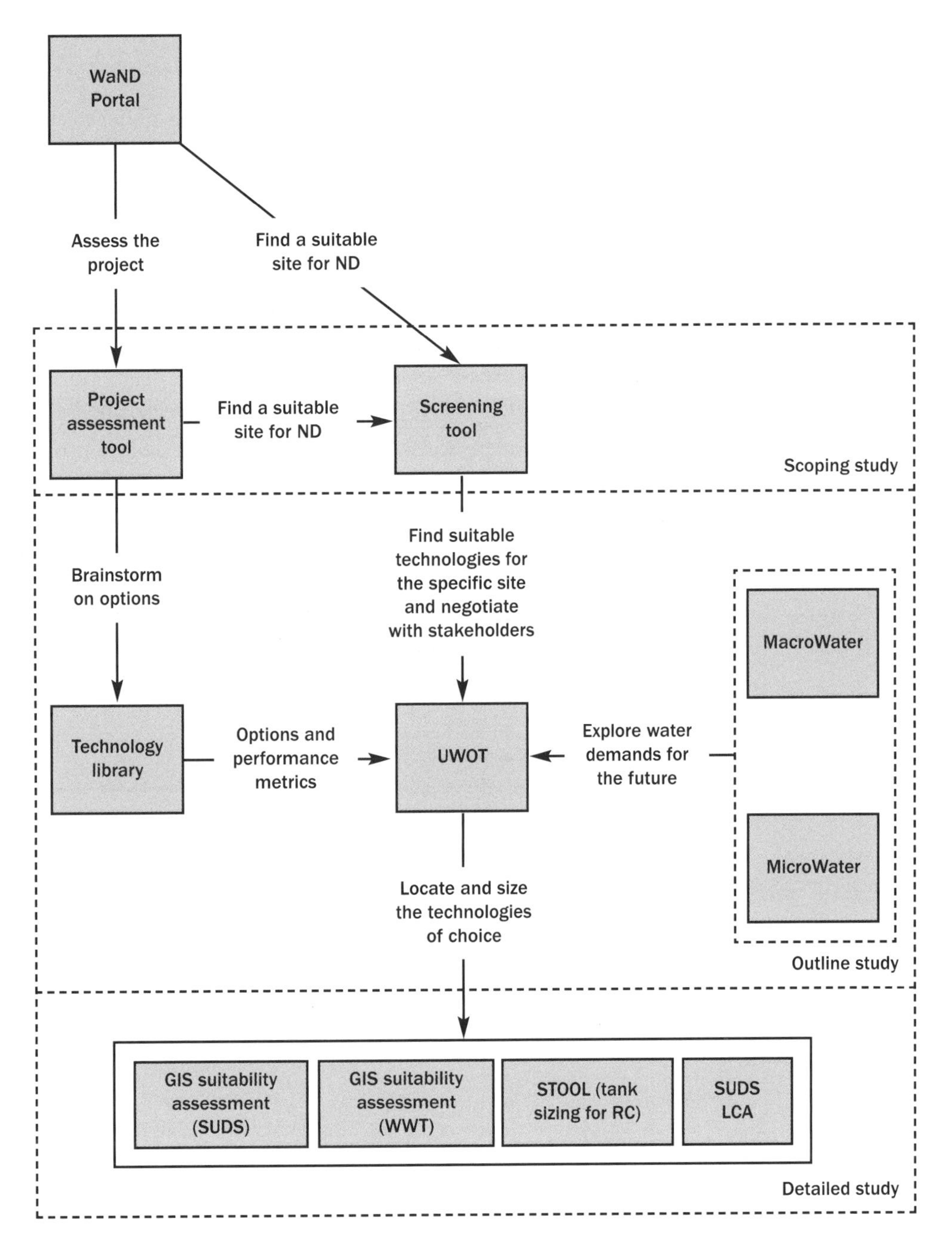

Figure 6.5 *WaND tools interaction*

Adaptive capacity and scenario planning

- water management strategies should be robust and perform well under a range of possible but initially uncertain future developments. This implies an increased use of scenario planning
- socio-economic and governance futures will have a significant influence on the choice, deployment and effect of urban water infrastructure. Horizon-scanning approaches, including long-term socio-economic scenario development, need to be integrated within decision-support tools to manage this exploration
- a flexible framework and scenario harmonisation and customisation approach could serve as a basis for more standardised scenario development and so assist in the comparability of results across disciplines and the development of a common vocabulary within the scientific community.

6.4 CONCLUSION

- the WaND project produced a comprehensive range of information, systems and tools to aid the decision maker
- the tools developed in the WaND research project are useful because they can either be applied in practice or have produced results that have helped establish the technologies and methods that are appropriate at different stages and scales of application
- the tools enable decisions from a macro to a micro scale and can assist with the decision making process as well as engagement with different sectors
- the undeveloped tools have great potential to be developed further by industry
- there is significant potential for a range of improvements in urban water management that could result from the context-aware use of a portfolio of technological infrastructure options.

6.5 FURTHER RESEARCH AND GUIDANCE

- **the tools need to be developed to contain a user-friendly interface**
- **decision-support tools cannot be used in isolation and cannot individually provide holistic sustainability evaluation.**

7 Final conclusions

This chapter:

- provides recommendations for sustainable water cycle management in new developments
- summarises messages for stakeholders.

7.1 RECOMMENDATIONS

This section provides several recommendations to achieve more sustainable water cycle management in new developments, bearing in mind the background and pressures discussed in Chapter 2, the issues of sustainability and the planning process presented in Chapter 3, the available technologies discussed in Chapter 4, the process of stakeholders engagement in Chapter 5 and the available decision-support tools described in Chapter 6.

Planning

Water is a vital component of any new development and historically it has been rather neglected in the planning process. It is recommended that water supply, wastewater disposal and surface water (stormwater) management are all fully considered early in the planning process as flood risk currently is. The Environment Agency water cycle study guidance report, the findings from the pilots on integrated urban drainage and surface water management plans as well as the Flood and Water Management Bill 2009 all need to be included in the planning process.

Water supply

Many parts of the UK are water-stressed, some seriously so, and any new development will have to deliver lower per capita consumption both in the new development and the existing adjacent developments. The water neutrality concept needs to be shown to be possible at scale (not just in very small developments).

Water-saving devices can considerably reduce water consumption and can be applied on a large scale. By adopting low-water appliances code level 3 for the Code for Sustainable Homes can be achieved. The appliances should be mandatory in all new homes and considered when retrofitting old housing stock, which is where the greatest contribution is likely to be experienced.

Rainwater harvesting is recommended in larger dwellings and in parts of the country that experience high rainfall. Carefully designed systems can save considerable volumes of water used for non-potable applications and provide benefit in terms of surface water management. The shortest payback periods are in schools and commercial buildings.

Greywater is more reliable all year round than rainwater harvesting but requires more extensive planning and design input, particularly with regard to treatment technologies. Single-house systems are not currently recommended, but systems are viable for multiple dwellings, blocks of flats, hotels or large public and commercial buildings. Consideration needs to be given to plumbing in a second, colour-coded system for greywater applications.

Water-saving technology and involvement on behaviours was thought to be the most successful approach to managing the short-term balance between water supply and demand.

Wastewater

As water consumption reduces wastewater generation will also reduce. Although increased specification of low-flush toilets is not seen as likely to increase the propensity for blockages in drains and small sewers, systems designs would need detailed consideration if ultra-low-flush toilets (ULFT's) are fitted en masse.

It is recommended that the new generation "waterless" devices are considered as they become available on the market.

Stormwater

It is strongly recommended that sustainable drainage systems (SUDS) are used as an important part of the surface water management in new developments. Consideration should also be given to capturing some of the stored stormwater for reuse in non-potable applications.

For SUDS to become a more universal solution local site conditions, opportunities and constraints need to be fully understood. They contribute to flood risk management but they are not the total answer. Modelling of the examples in WaND shows that even with extensive use of SUDS, greenfield hydrological responses cannot be achieved post-development.

It is important that maintenance requirements and more specifically the adoption of SUDS schemes are clearly understood by all parties to ensure smooth functioning post development.

Energy

Sustainable water cycle management (SWCM) in new developments can make an important contribution to reducing carbon emissions, but regulation of water service providers also needs to ensure that reduced carbon emissions are derived from "source to tap". Stronger links need to be made between the water cycle and energy to encourage water saving.

Community

The majority of householders have little appreciation of the seriousness of the water situation in the UK. It is recommended that there is a concerted campaign to make householders understand the value of water and how each household can reduce its water footprint. This needs to be much more than information provided on water bills and the messages need to be delivered over a period of years to encourage changes in behaviour.

No evidence has been found to suggest that SWCM approaches will compromise public health. Where dual supplies are fitted for potable and non-potable applications it is recommended that different diameters, colour coding or appropriate marking of the pipework is used to differentiate the use.

The effect of climate change, which is likely to increase the frequency of extreme weathers patterns where parts of the country may experience droughts one year (summer 2006) and flooding in the next (summer 2007), makes this engagement even more challenging.

Costs

Rainwater harvesting can deliver a reasonable payback on large individual dwellings public buildings, such as schools, and medium and large commercial buildings.

Greywater recycling is most appropriate for buildings such as hotels and blocks of flats as well as some public and commercial buildings.

Evidence suggests that appropriate SUDS schemes have lower capital and whole-life costs than traditional systems (with clarity of who is responsible for maintenance).

Now individual households can do a great deal in terms of fittings and low-use appliances, however, consideration needs to be given as to how a water-efficiency market can be pump-primed or include incentives to ensure greater uptake. There may be lessons to be learnt from the energy sector in terms of change towards energy-efficient lighting and loft insulation.

When considering SWCM there are a range of benefits to consider. Carbon and cost are often the main drivers but the wider benefits should be considered. SWCM helps conserve water while also adding value to the community by creating environments that encourage wildlife and recreation, creating spaces for people to interact.

7.2 KEY MESSAGES

A summary of the key messages discussed in this guidance document is presented in Table 7.1.

Table 7.1 *Key messages for stakeholders*

Target audience	Key messages
Central government and regulators	• provide consistent messages across all government departments and agencies • produce mandatory water targets in the Code for Sustainable Homes • establish exemplars of good practice (eco-towns are a good opportunity) • develop SWCM incorporation into water cycle studies and ensure this is embedded in the planning process • raise public awareness on water issues • produce non-potable water quality standards.
Local government and planners	• become more prominent in promoting sustainable water cycle management • provide education (and resources) for all aspects of sustainable water cycle management • ensure SWCM is considered early in the planning process • use new decision-support tools to prioritise and compare sites for development with respect to water, wastewater and surface water • incorporate sustainability criteria into decision making • reassure stakeholders on public health matters arising from SWCM approaches.
Developers, architects and consultants	• be willing to specify and trial new SWCM approaches • volunteer to plan, design and build exemplar SWCM developments • accept that some costs may initially be higher • use new decision-support tools to estimate water and energy savings • undertake health impact assessments for new schemes to reassure all stakeholders.
Water service providers	• become more prominent in promoting and supporting SWCM • provide education (and resources) on water cycle management • provide clear water efficiency messages on bills and all communications with customers • involve local authorities with SWCM early in the planning process • be willing to specify and trial new approaches • volunteer to take part in exemplar developments • accept that some costs may initially be higher • involve customers regularly and consistently on both water efficiency and SWCM • use new decision-support tools to estimate water and energy savings • use urban water futures to plan for future water needs.

References

ASHLEY, R, BLACKWOOD, D, BUTLER, D, and JOWITT, P (2004)
Sustainable water services. A procedural guide
IWA Publishing, London (ISBN: 1-84339-065-5)

ASHLEY, R, BUTLER, D, HURLEY, L and MEMON, F (2008)
"Criteria and indicators in delivering sustainable water systems: from Sustainable Water Asset Resource Decisions (SWARD) to Water cycle management for New Developments (WaND)"
In: *DayWater: an adaptive decision support system for urban stormwater management,* Daniel R Thevenot (ed), IWA Publishing, London (ISBN: 978-1-84339-160-9)

AZAR, C, HOLMBERG, J and LIDGREN, K (1996)
"Socio-ecological indicators for sustainability"
Ecological Economics, C J Cleveland (ed), vol 18, **2**, Elsevier BV, pp 89–112

BALKEMA, A J (2003)
Sustainable wastewater treatment: developing a methodology and selecting promising systems
PhD Thesis, University of Eindhoven

BERKHOUT, F, HERTIN, J and JORDAN, A (2002)
"Socio-economic futures in climate change impact assessment: using scenarios as 'learning machines'."
Global Environmental Change, 12, **2**, Elsevier BV, pp 83–95

BREWER, D, BROWN, R and STANFIELD, G (2001)
Rainwater and greywater in buildings – project report and case studies
BSRIA publications. Technical Note TN 7/2001

BROWN, R R, SHARP, L and ASHLEY, R (2005)
"Implementation impediments to institutionalising the practice of sustainable urban water management"
In: *Proc 10th int conf on urban drainage, Copenhagen, Denmark*, 21–26 August 2005

BRUNNER, N, and STARKL, M (2004)
Decision aid systems for evaluating sustainability: a critical survey
Environmental Impact Assessment Review, 24, pp 441–469

BUTLER, D and DAVIES, J W (2004)
Urban drainage, 2nd edition
Spon Press, London

BUTLER, D, KOKKALIDOU, A and MAKROPOULOS, C K (2005)
"Supporting the siting of new urban developments for integrated urban water resource management"
In: *Integrated urban water resources management*, P Hlavinek and T Kukharchyk (eds), NATO Scientific Series, Springer, pp 19–34

BUTLER, D and MAKROPOULOS, C (2006)
Water related infrastructure for sustainable communities. Technological options and scenarios for infrastructure systems
Science Report SC05002501, Environment Agency, Bristol (ISBN: 1-84432-611-X).
Available from: <http://publications.environment-agency.gov.uk>

BUTLER, D and PARKINSON, J (1997)
"Towards sustainable urban drainage"
Water Science and Technology, 35, **9**, Elsevier BV, pp 53–63

CLG (2005)
Planning Policy Statement (PPS) 1: *Delivering sustainable development*
Department of Communities and Local Government, HMSO London, January 2005

CLG (2006)
Planning Policy Statement (PPS) 25: *Development and flood risk*
Department of Communities and Local Government, HMSO London, December 2006

CLG (2006)
Building a greener future: towards zero carbon development
06 SCDD 04276, Crown Copyright, London. Available from:
<http://www.communities.gov.uk/documents/planningandbuilding/pdf/153125.pdf>

CLG (2008)
Planning Policy Statement (PPS) 25: *Development and flood risk – a practical guide*
Department of Communities and Local Government, HMSO London, June 2008.
Available from: <http://www.communities.gov.uk/documents/planningandbuilding/pdf/pps25guideupdate.pdf>

CLG (2009)
Code for sustainable homes – technical guide
Version 2, code: 08BD05840, Department for Communities and Local Government, London (ISBN: 978-1-85946-330-7). Available from:
<http://www.planningportal.gov.uk/uploads/code_for_sustainable_homes_techguide.pdf>

DEFRA (2008a)
Future water
Cm 7317, Department for Environment Food and Rural Affairs. HM Government, London. Available from:
<http://www.defra.gov.uk/environment/water/strategy/pdf/future-water.pdf>

DEFRA (2008b)
Sustainable development – shared UK principles of sustainable development
Department for Environment Food and Rural Affairs, London. Available from:
<http://www.defra.gov.uk/sustainable/government/what/principles.htm>

DEFRA (2008c)
Statutory social and environmental guidance to the Water Services Regulation Authority (Ofwat)
Department for Environment Food and Rural Affairs, London. Available from:
<www.defra.gov.uk/corporate/consult/ofwat-guidance/>

DEFRA (2009a)
Flood and Water Management Bill
Department for Environment Food and Rural Affairs, London. Go to:
<http://www.parliament.uk/parliamentary_committees/environment__food_and_rural_affa irs/efra_draft_flood_and_water_bill.cfm>

DEFRA (2009b)
Nation of water wasters – new campaign to save 20 litres a day
Department for Environment Food and Rural Affairs, London. Go to:
<http://www.defra.gov.uk/News/2009/090924a.htm>

DIXON, J and SHARP, L (2007)
"Collaborative research in sustainable water management: issues of interdisciplinary"
SR Interdisciplinary science reviews, Maney, Leeds, 32, **3**, pp 221–232

DoENI (2006)
Planning Policy Statement (PPS) 15: *Planning and flood risk*
Department of the Environment Northern Ireland, Belfast

DTI (2006)
Energy: its impact on the environment and society
Department of Trade and Industry, London. Available from:
<http://www.berr.gov.uk/files/file32546.pdf>

DUFFY, A, JEFFERIES, C, BLACKWOOD, D, WADDELL, G, SHANKS, G and WATKINS, A (2008)
"A cost comparison of traditional drainage and SUDS in Scotland"
Water Science and Technology, vol 57, **9**, Elsevier BV, pp 1451–1459

EA (1999)
Water consumption and conservation in buildings. Review of water conservation measures
Environment Agency, Bristol

EA (2001a)
Conserving water in buildings "fast cards"
Environment Agency, Bristol

EA (2001b)
A scenario approach to water demand forecasting.
National Water Demand Management Centre Reports, Environment Agency, Bristol

EA (2006)
A guide for developers – practical advice on adding value to your site. Building a better environment
Environment Agency, Bristol

EA (2007a)
Summary report – towards water neutrality in the Thames Gateway
Science report: SC060100/SR3, Environment Agency, Bristol. Available from:
<http://www.environment-agency.gov.uk/subjects/waterres/287169/1917628/?lang=_e>

EA (2007b)
Conserving water in buildings. A practical guide
Environment Agency, Bristol

EA (2007c)
Water demand management
Bulletin 84, Environment Agency, Bristol

EA (2008a)
Harvesting rainwater for domestic uses: an information guide
Environment Agency, Bristol. Available from: <http://www.environment-agency.gov.uk/commondata/acrobat/geho0108bnpnee_809069.pdf>

EA (2008b)
Greywater: an information guide
GEHO0408BNWQ-E-E, Environment Agency, Bristol. Available from: <http://www.environment-agency.gov.uk/commondata/acrobat/geho0408bnwqee_2033772.pdf>

EA (2009a)
Water cycle study guidance
GEO0109BPFF-E-E, Halcrow and Environment Agency report, Environment Agency, Bristol

EA (2009b)
Water for people and the environment – water resources strategy for England and Wales
GEHO0309BPKX-E-P, Environment Agency, Bristol. Available from: <http://publications.environment-agency.gov.uk/pdf/GEHO0309BPKX-E-E.pdf>

ELLIS, J B, SHUTES, R B E and REVITT, M D (2003a)
Constructed wetlands and links with sustainable drainage systems
R&D Report P2-159/TR1, Environment Agency, Bristol

ELLIS, J B, SHUTES, R B E and REVITT, M D (2003b)
Guidance manual for constructed wetlands
R&D Report P2-159/TR2, Environment Agency, Bristol

ENVIROWISE (1996)
Saving money through waste minimisation: reducing water use
GG26R, Evirowise, Oxfordshire. Go to: <http://www.envirowise.gov.uk>

EVANS, E, ASHLEY, A, HALL, J, PENNING-ROWSELL, E, SAUL, A, SAYER, P, THOME, C and WATKINSON, A (2004)
"Foresight future flooding"
Scientific summary: Volume 1 – Future risks and their drivers, Office of Science and Technology, London

FEWKES, A (2006)
"The technology, design and utility of rainwater catchment systems"
Water demand management, D Butler and F A Memon (eds), IWA Publishing (ISBN: 1-84339-078-7), pp 27–61

FOXON, T J, MCILKENNY, G, GILMOUR, D, OLTEAN-DUMBRAVA, C, SOUTER, N, ASHLEY, R, BUTLER, D, PEARSON, P, JOWITT, P and MOIR, J (2002)
"Sustainability criteria for decision support in the UK water industry"
Environmental Planning and Management, 45, **2**, Routledge, UK, pp 285–301

GALLOPÍN, G, HAMMOND, A, RASKIN, P and SWART, R (1997)
Branch points: global scenarios and human choice
PoleStar Series, Stockholm Environment Institute, Stockholm

GRANT, N (2002)
Water conservation products – a preliminary review
Elemental Solutions, Hereford. Available from:
<http://www.elementalsolutions.co.uk/publications.htm>

GREEN, C and WILSON, T (2004)
Assessing the benefits of reducing the risk of flooding from sewers
Flood Hazard Research Centre, Middlesex University, UK. Go to:
<http://www.ofwat.gov.uk/pricereview/pr04/pr04phase3/rpt_com_reducerisksewrfld.pdf>

GRIGGS, J and BURNS, J (2008)
Water efficiency in new homes. An introductory guide for housebuilders
IHS BRE Press on behalf of the NHBC Foundation (ISBN: 978-1-84806-099-9). Available from: <http://www.nhbcfoundation.org>

GROSS, A, SHMUELI, O, RONEN, Z and RAVEH, E (2007)
"Recycled vertical flow constructed wetland (RVFCW) – a novel method of recycling greywater for irrigation in small communities and households"
Chemosphere, 66, **5**, Elsevier BV, pp 916–923

HATT, B, DELETIC, A and FLETCHER, T (2004)
Integrated stormwater treatment and reuse systems
Technical Report 04/1, Inventory of Australian practice. Available from:
<http://iswr.eng.monash.edu.au/research/projects/stormwater/hattwsud04.pdf>

HESPANHOL, I. and PROST, A. M. E. (1994)
"WHO guidelines and national standards for reuse and water quality"
Water Research, 28, **1**, pp 119–127

HEUGENS, P and VAN OOSTERHOUT, J (2001)
"To boldly go where no man has gone before: integrating cognitive and physical features in scenario studies"
Futures, 33, **10**, pp 861–872

HOLMAN, I P (2002)
Regional Climate Change Impact and Response Studies in East Anglia and North West England (RegIS)
Cranfield University, Bedfordshire, UK. Available from:
<https://www.cranfield.ac.uk/sas/naturalresources/research/projects/regis.jsp>

HURLEY, L, ASHLEY, R and MOUNCE, S (2008)
"Addressing practical problems in sustainability assessment frameworks"
Engineering Sustainability, 161 (ES1), pp 23–30

HURLEY, L, MOUNCE, S, ASHLEY, R and MAKROPOULOS, C K (2007)
"Support for more sustainable decision making in urban water management"
In: *Proc Water management challenges in global change, CCWI 2007 and SUWM 2007 conf, De Montford, Leicester, UK, 3–5 September 2007*, pp 577–584

JACKSON, T (2005)
Motivating sustainable behaviour: A review of evidence on consumer behaviour and behavioural change
Centre for Environment Strategy, University of Surrey, Guildford

KELLAGHER, R B and LAUCHLAN, C S (2005)
Use of SUDS in high density developments
Report SR 666, version 3.0, HR Wallingford

KELLAGHER, R B and MANEIRO FRANCO, E (2007)
Rainfall collection and use in developments; benefits for yield and stormwater control
WaND Briefing Note 19, HR Wallingford

KNAMILLER, C, SEFTON, C, SHARP, E and MEDD, W (2007)
"Water in everyday use: a study on water-using technologies and the water user in Essex"
In: *Proc Water management challenges in global change, CCWI 2007 and SUWM 2007 conf, De Montford, Leicester, UK, 3–5 September 2007*, pp 329–336

KNOPS, G, PIDOU, M, JUDD, S J and JEFFERSON, B (2007)
"The impact of household products on biological systems for water reuse"
In: *Proc 6th IWA Wastewater reclamation and reuse for sustainability, Antwerp, Belgium, 9–12 October 2007*

LEGGETT, D J, BROWN, R, BREWER, D, STANFIELD, G and HOLLIDAY, E (2001)
Rainwater and greywater use in buildings. Best practice guidance
C539, CIRIA, London (ISBN: 978-0-86017-539-1)

LEGGETT, D J, BROWN, R, STANFIELD, G and HOLLIDAY, E (2001)
Rainwater and greywater use in buildings. Decision-making for water conservation
PR80, CIRIA, London (ISBN: 978-0-86017-880-4)

LI PUMA, G, TOEPFER, B and GORA, A (2007)
"Photocatalytic oxidation of multicomponent systems of herbicides: Scale-up of laboratory kinetics rate data to plant scale"
Catalysis Today, 124, **3–4**, Elsevier BV, pp 124–132

LIU, S, MAKROPOULOS, C K, MEMON, F A and BUTLER, D (2007)
"A storage tank sizing tool for grey water recycling systems"
In: *Proc Water management challenges in global change, CCWI 2007 and SUWM 2007 conf, De Montford, Leicester, UK, 3–5 September 2007*

MAKROPOULOS, C K, MEMON, F A, SHIRLEY-SMITH, C and BUTLER D (2008)
"Futures: an exploration of scenarios for sustainable urban water management"
Water Policy, 10, **4**, pp 345–373

MAKROPOULOS, C, K NATSIS, K, LIU, S, MITTAS, K and BUTLER, D (2008)
"Decision support for sustainable option selection in integrated urban water management"
Environmental Modelling & Software, vol 23, **12**, Elsevier BV, pp 1448–1460

MARKET TRANSFORMATION PROGRAMME (MTP) (2007)
Rainwater and grey water: review of water quality standards and recommendations for the UK
Department for Environment Food and Rural Affairs, London. Go to: <http://efficient-products.defra.gov.uk/cms/market-transformation-programme/>

MELBOURNE WATER (2005)
Water sensitive urban design-engineering procedures: stormwater
CSIRO Publishing, Australia (ISBN: 978-0-64309-092-7)

MEMON, F A and BUTLER, D (2006)
"Water consumption trends and demand forecasting techniques"
Water demand management, D Butler and F A Memon (eds), IWA Publishing
(ISBN: 1-84339-078-7), pp 1–26

MEMON, F A, BUTLER, D, HAN, W, LIU, S, MAKROPOULOS, C K, AVERY, L, and PIDOU, M (2005)
"Economic assessment tool for greywater recycling systems"
Engineering Sustainability, 158, **ES3**, Institution of Civil Engineers, pp 155–161

MEMON, F A, ZHENG, Z, BUTLER, D, SHIRLEY-SMITH, C, LIU, S MAKROPOULOS, C K, and AVERY, L (2007a)
"Life cycle impact assessment of greywater treatment technologies for new developments"
Environmental Monitoring and Assessment, 129, **1–3**, Springer, Netherlands, pp 27–35

MEMON, F A, FIDAR, A, LITTLEWOOD, K, BUTLER, D, MAKROPOULOS C K and LIU, S (2007b)
"A performance investigation of small-bore sewers"
Water Science and Technology, 55, **4**, Elsevier BV, pp 85–91

MUSTOW, S and GREY, R (1997)
Greywater and rainwater systems: recommended UK requirements
Report 13034/2, BSRIA Ltd, Bracknell, UK

ODPM (2003)
Millennium villages and sustainable communities: final report
Office of the Deputy Prime Minister, London

OFWAT (2007)
Strategic direction statements. A letter to all managing directors of water and sewage companies and water companies only
MD233. Available from: <http://www.ofwat.gov.uk/pricereview/pr09phase1/pr09phase1letters/ltr_md223_stratdirestate>

PIDOU, M, MEMON, F A, STEPHENSON, T, JEFFERSON, B and JEFFREY, P (2007)
"Greywater recycling: a review of treatment options and applications"
Engineering Sustainability, 160, **ES3**, Institution of Civil Engineers, pp 119–131

PRATT, C, WILSON, S and COOPER, P (2002)
Source control using constructed pervious surfaces: Hydraulic, structural and water quality performance issues
C582, CIRIA, London (ISBN: 978-0-86017-582-7)

QUEENSLAND GOVERNMENT (2003)
Onsite sewerage facilities. Guidelines for the use and disposal of greywater in unsewered areas
Queensland Government, Local Government and Planning, Brisbane, Queensland, Australia

RASKIN, P, BANURI, T, GALLOPÍN, G, GUTMAN, P, HAMMOND, A, KATES, R and SWART, R (2002)
Great transition: the promise and lure of the times ahead
PoleStar Series, Stockholm Environment Institute, Stockholm

RACHWAL, A J and HOLT, D (2008)
Urban rainwater harvesting and water reuse: a review of potential benefits and current UK best practices
FR/G0006, a foundation for water research guide. Available from:
<http://www.fwr.org/sewerage.htm>

SAKELLARI, I, MAKROPOULOS, C K, BUTLER, D and MEMON, F A (2005)
"Modelling sustainable urban water management options"
Engineering Sustainability, 158, **ES3**, Institution of Civil Engineers, pp 143–153

SAMUELS, P, WOODS-BALLARD, B, HUTCHINGS, C, FELGATE, J, MOBBS, P, ELLIOT, C and BROOK, D (2006)
Sustainable water management in land-use planning
C630, CIRIA, London (ISBN: 978-0-86017-630-5)

SCOTTISH EXECUTIVE (2004)
Scottish Planning Policy (SPP) 7: *Planning and flooding*
The Scottish Office, Edinburgh (ISBN: 0-7559-2439-8)

SEFTON, C and SHARP, L (2005)
Evaluating pro-environmental messages: the promotion of water efficiency in England and Wales
WaND (WP4/WP14) Briefing Note 16 (available in the CD portal)

SEFTON, C and SHARP, L (2007)
"Why we should celebrate water: Recommendations for engaging the public in sustainable water management"
In: *Proc Water management challenges in global change, CCWI 2007 and SUWM 2007 conf, De Montford, Leicester, UK, 3–5 September 2007* pp 681–688

SHAFFER, P, ELLIOT, C, REED, J, HOMES, J and WARD, M (2004)
Model agreements for sustainable water management systems. Model agreements for SUDS
C625, CIRIA, London (ISBN: 978-0-86017-625-1)

SHAFFER, P, ELLIOTT, C, REED, J, HOLMES, J and WARD, M (2004)
Model agreements for sustainable water management systems. Model agreement for rainwater and greywater use systems
C626, CIRIA, London (ISBN: 978-0-86017-626-8)

SHARP, L (2006)
"Water demand management in England and Wales: constructions of the domestic water user"
Journal of Environmental Planning and Management, vol 49, **6**, pp 869–889

SIM, P, MCDONALD, A, PARSONS, J and REES, P (2007)
Revised options for UK domestic water reduction: a review
WaND Briefing Note 28, working paper 07/04, University of Leeds, UK

STERN, N (2005)
Stern review on the economics of climate change
HM Treasury, London. Available from:
<http://www.hm-treasury.gov.uk/sternreview_index.htm>

SURENDRAN, S and WHEATLEY, A D (1998)
"Greywater reclamation for non-portable reuse"
Water and Environment Journal, CIWEM, vol 12, **6**, pp 406 – 413

TAJIMA, A (2005)
"The behaviour of the pathogenic microbes in the treated wastewater reuse system and the establishment of the new technical standard for the reuse of treated wastewater"
In: *Proc IWA specialty conf on wastewater reclamation and reuse for sustainability, 8–11 November, Jeju, Korea*

TURBAN, E (2007)
Decision Support Systems and Intelligent Systems, 8/E
Prentice Hall (ISBN: 978-0-13046-106-3)

UDALE-CLARKE, H and KELLAGHER, R (2007)
Elvetham Heath case study – testing of sustainability measures for stormwater
Report SR 684, HR Wallingford

UKWIR (2003a)
Effect of climate change on river flows and groundwater recharge UKCIP 02 scenarios
03/CL/04/2, UK Water Industry Research, London (ISBN: 1-84057-286-8)

UKWIR (2003b)
Quantification of the savings, costs and benefits of water efficiency
03/WR/25/1, UK Water Industry Research, London (ISBN: 1-84057-291-4)

UKWIR (2005)
Framework for developing water reuse criteria with reference to drinking water supplies
05/WR/29/1, UK Water Industry Research, London (ISBN: 1-84057-368-6)

UKWIR (2006a)
Sustainability of water efficiency
06/WR/25/2, UK Water Industry Research, London (ISBN: 1-84057-403-8)

UKWIR (2006b)
21st century sewerage design:summary report
06/WM/07/6, UK Water Industry Research, London (ISBN: 1 84057-426-7)

UKWIR (2007)
A framework for valuing the options for managing water demand
07/WR/25/3, UK Water Industry Research, London (ISBN: 1-84057-462-3)

UPHAM, P (2000)
"An assessment of the natural step theory of sustainability"
Journal of Cleaner Production, vol 8, **6**, Elsevier BV, pp 445–454

USEPA (2004)
Guidelines for water reuse
Report EPA/625/R-04/108, U S Environmental Protection Agency, Washington, DC, USA

VOADEN, D (2008)
LCA of sustainable drainage systems
BEng project, University of Exeter

WATERWISE (2008)
Evidence base for large scale water efficiency in homes
WaterWise, UK. Go to:
<http://www.waterwise.org.uk/images/site/Policy/evidence_base/evidence%20base%20for%20large-scale%20water%20efficiency%20in%20homes,%20waterwise,%20october%202008.pdf>

WATERWISE (2009)
Reducing waste water in the UK
WaterWise, UK. Go to: <http://www.waterwise.org.uk/reducing_water_wastage_in_the_uk/the_facts/the_facts_about_saving_water.html>

WATERWISE (2009)
The water and energy implications of bathing and showering behaviours and technologies
WaterWise, UK. Go to:
<http://www.waterwise.org.uk/images/site/Research/final%20water%20and%20energy%20implications%20of%20personal%20bathing%20-%20for%20est%20apr%2009.pdf>

WELSH ASSEMBLY GOVERNMENT (2004)
Technical Advice Note (TAN) 15: *Development and flood risk*
Planning Policy Wales, National Assembly for Wales, Cardiff (ISBN: 0-75043-501-1)

WHO (1964)
World Health Organisation Constitution
World Health Organization, Geneva. Available from:
<http://www.who.int/water_sanitation_health/wastewater/gsuww/en/index.html>

WHO ECHP (1999)
Health Impact Assessment: main concepts and suggested approach
Gothenbury consensus paper, World Health Organization European Centre for Health Policy

WHO (2006)
"Guidelines for the safe use of wastewater, excreta and greywater"
Volume II: Wastewater use in agriculture, World Health Organization, Geneva. Available from:
<http://www.who.int/water_sanitation_health/wastewater/gsuww/en/index.html>

WILSON, S (2008
Structural design of modular geocellular drainage tanks
C680, CIRIA, London (ISBN: 978-0-86017-680-0)

WILSON, S, BRAY, R and COOPER, P (2004)
Sustainable drainage systems: Hydraulic, structural and water quality advice
C609B, CIRIA, London (ISBN: 978-0-86017-609-1)

WOODS-BALLARD, B (2007)
The SUDS manual
C697, CIRIA, London (ISBN: 978-0-86017-697-8)

WOODS-BALLARD B, KELLAGHER R et al (2007)
Site handbook for the construction of SUDS
C698, CIRIA, London (ISBN: 978-0-86017-698-5)

WOODS-BALLARD, B, KELLAGHER, R, MARTIN, P, JEFFERIES, C, BRAY, B and
WORLD COMMISSION ON ENVIRONMENT AND DEVELOPMENT (1987)
Our common future (The Brundtland Report)
Oxford University Press, Oxford (ISBN: 0-19282-080-X)

WWUK (2008)
GROW: Green Roof Recycling System
Water Works UK Ltd. Go to:
<http://www.wwuk.co.uk/grow.htm>

Acts and Bills

Climate Change Act 2008

Climate Change (Scotland) Act 2009

Flood and Water Management Bill 2009

Water Industry Act 1991 and the updated Water Industry Act 1999

Water Act 2003

British Standards

BS 8515:2009 *Rainwater harvesting systems. Code of practice*

Regulations

Building Standards (Scotland) Regulations 1990

Water Supply (Water Fittings) Regulations 1999

Water Supply (Water Quality) Regulations 2001 (Amendment) Regulations 2007

A1 The WaND research project and portal

WaND RESEARCH PROJECT

Water cycle management in new developments (WaND) was a multi-disciplinary EPSRC research project aimed at exploring and developing sustainable approaches to water management in new developments and minimising the potential effect on the water cycle. The WaND project was developed to manage the challenges faced by developers, planners, water service provider (WSPs) and the government on how to provide the necessary expansion in housing stock in a sustainable way. A particular aspect of this is the potential effect on the water cycle, in terms of water supply requirements, wastewater management and stormwater management. WaND aimed to create the provision of tools and guidelines for project design, adoption and management for the delivery of sustainable water management at the local level.

The work packages that formed the WaND research	Title
Work package 1	Water supply
Work package 2	Stormwater
Work package 3	Wastewater collection
Work package 4	Social and economic aspects
Work package 5	Environmental health aspects
Work package 6	The toolbox
Work package 7	On-site soil and wastewater treatment options for new developments technologies, economics and management.
Work package 8	Flexible frameworks to include sustainable perspectives into urban water decision making systems (FlexiFrame)
Work package 9	Integrated modelling and advanced decision support
Work package 10	SUDS performance at development scale – interactions with rainfall and groundwater
Work package 11	Strategic project and planning guidance on cross-sectoral and scale-up issues for sustainable development
Work package 12	Household data for water, land and waste management
Work package 14	Innovations and risks

Note: there was no work package 13

Work packages that formed the WaND research where mainly focused on the technical aspects (water supply, wastewater collection, storm drainage and SUDS). However a range of issues were included in the other work packages such as social, planning, economic and health issues.

The project was co-ordinated by Professor David Butler of Exeter University and included a substantial input from the universities of Bradford, Cranfield, Leeds, Sheffield and Wales (Aberyswyth) plus the Centre for Ecology and Hydrology (CEH), HR Wallingford and Water Resource Centre (WRc).

WaND PORTAL

The WaND portal provides further details of the project, allows the user to explore the various work packages and provides access to some of the deliverables. The portal is accessible through the CD-Rom included with this guidance. It uses a GUI (graphic user interface) that guides you through different aspects of the WaND project including sustainability, water management technologies, tools and stakeholder engagement. It also provides details of the case studies used in the WaND research.

A2 Further discussion on sustainability

The question "what is sustainable?" largely depends on the stakeholder asking the question, the context, where, when and under what circumstances is the question being posed? The widely accepted ingredients of sustainability (society, the environment and the economy) are too broad to allow for uniform interpretation across the range of stakeholders and some tensions are likely to be encountered.

Broad concepts of development requirements, social well-being, equity, environmental protection and economic growth are open to interpretation at the societal and individual level and possibilities for serious, long-term forecasts are limited (an approach to thinking about the future is explained in Chapter 3).

It has been suggested that sustainable development when considered as an outcome may be difficult to define (Ashley *et al*, 2008). However, when considered as a process of learning and communication, a journey that is constantly reviewed and adapted from new knowledge and information, more sustainable development becomes a realistic target (Box A2.1).

The European Commission's Joint Research Centre (JRC), when exploring the practical application of sustainability, suggests that sustainable development requires a focus on interaction between multiple disciplines including planners and engineers, but also social scientists. The innovative inclusion of a broad range of viewpoints and an acceptance that, apart from technological change, behavioural patterns need to also change to assist resource reduction and to move towards sustainability.

There is a growing acceptance that multi-disciplinary, multi-actor approaches can be more effective when set up as long-term collaborations rather than ad hoc teams built on a project-by-project basis. This is because the development of a common language, which will allow the sustainability debate to take place, is a long and difficult process (Dixon and Sharp, 2007).

Researchers, policy-makers and decisions makers are demanding focus on adaptive management as a strategy for managing these issues. This was central to the WaND research project.

The delivery of sustainable development requires an innovative and adaptive approach to decision making. At a policy level, the benefits of sustainable development may be irrefutable, however, the translation into practice is not straightforward. Sustainable development requires integration between different professionals as well as those affected by the decision. In practice, a lack of communication and sharing information and knowledge across different professions and stakeholders, leads to using only tried and tested practices (Brown *et al*, 2005) and over-reliance on monetary evaluation to determine appropriate solutions.

Box A2.1 *The journey ahead: into the unknown (courtesy D Butler)*

Charles Lindbergh was the first man to fly across the Atlantic. He did so in a plane, called the Spirit of St Louis, which apart from its small size and fragile appearance, is most striking because it has no front window. Its only windows are on the side. This was because all the spare space available was filled with fuel tanks. Lindbergh could not see what lay ahead but only see what was below him and to the sides. He had a compass to give him a broad heading, but very few other navigational aids. This meant he could only know where he had been (the past) and where he was at the moment (the present) but he could not know or see exactly where he was going (the future).

And so it is with the journey towards sustainability – into the unknown and the unknowable. There is the compass of sustainability principles (society, economy and environment) but there is no clear idea of where the destination is. However, as history shows, Lindbergh was successful in navigating purely by the past and present and extrapolating intelligently into the future. So the journey of discovery along the route to sustainability is as important as the final destination. And the most important thing of all is to actually start the journey accepting some wrong turnings may be made along the way.

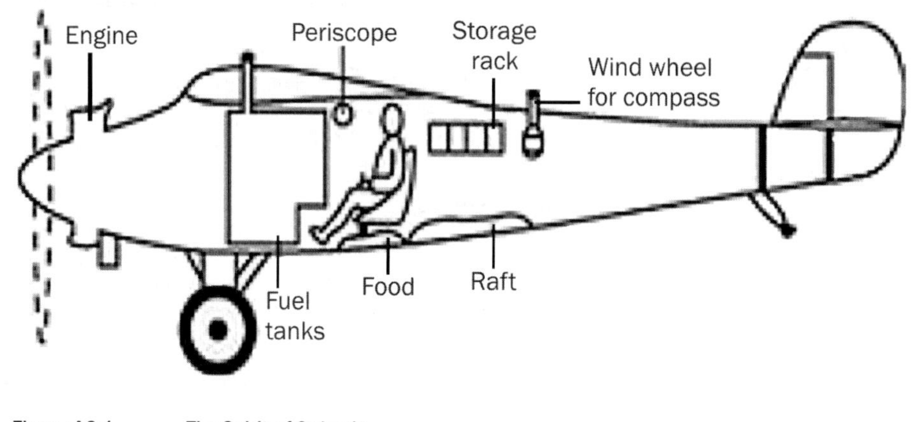

Figure A2.1 *The Spirit of St Louis*

A3 Water cycle studies

A3.1 WATER CYCLE STUDIES

Water cycle studies adopt an integrated approach to water infrastructure planning, which helps stakeholders meet the challenge issues of urban growth. The close co-ordination between main stakeholders and organisation builds confidence between parties and leads to a successful WCS.

A water cycle study is:

- a general method for ensuring that the most sustainable water infrastructure is provided where and when it is needed
- a way of ensuring that urban planning makes best use of environmental capacity, adapts to environmental constraints and optimises the use of environmental opportunities
- a way of ensuring that all stakeholders have their say
- helps make a more integrated planning decisions by bringing together all the available information and knowledge.

A WCS is broken down into three stages, to coincide with the planning process (Table A3.1).

Table A3.1 *Stages of the water cycle studies*

Stage	Description
Scoping study	Undertaken at an early stage in the preparation of the local development framework. The purpose of a scoping study is to identify at a very high level: ♦ what development is proposed ♦ what elements of the water cycle may be affected by the scale of development ♦ which authorities are responsible for the various elements of the water cycle ♦ what studies have already been carried out. The scoping study also identifies how the WCS fit in relation to other planning processes.
Outline study	Works alongside the preparation of the local development framework (LDF) core strategy. The outline study will: ♦ identify any significant environmental capacity constraints to development ♦ identify infrastructure constraints to development ♦ identify if major new infrastructure should be planned to allow development ♦ assist local authorities to identify where development is most feasible (or sustainable) with respect to water infrastructure ♦ provide the evidence base for the local planning authorities core strategy. If no significant constraints are identified, it should allow objections by the Environment Agency and other regulators to be lifted subject to agreement on how any required measures will be funded and provided. The outline study may identify more detailed assessments that need to be undertaken before new developments can be finally agreed. These will need to be assessed during a detailed WCS.
Detailed water cycle study	Works alongside the latter stages of the LDF process. This leads to a **water cycle strategy** that: ♦ identifies what water cycle infrastructure is required and where it is needed ♦ identifies who is responsible for providing the infrastructure and when it has to be provided by ♦ provides guidance for local authorities and developers on site specific infrastructure requirements (eg SUDS requirements). This stage may not be necessary if the previous stages have not identified any significant environmental risks or major infrastructure needs.

Adopting the strategy is then monitored as part of the planning application process. The compliance with the water cycle strategy can be included as part of the annual monitoring report, or as a periodic review with the water companies and the Environment Agency.

A water cycle study needs to answer different questions at different stages of the planning process. The flow chart (Figure A3.1) highlights questions that need to be considered in the different stages of a water cycle study in view of the overall planning process.

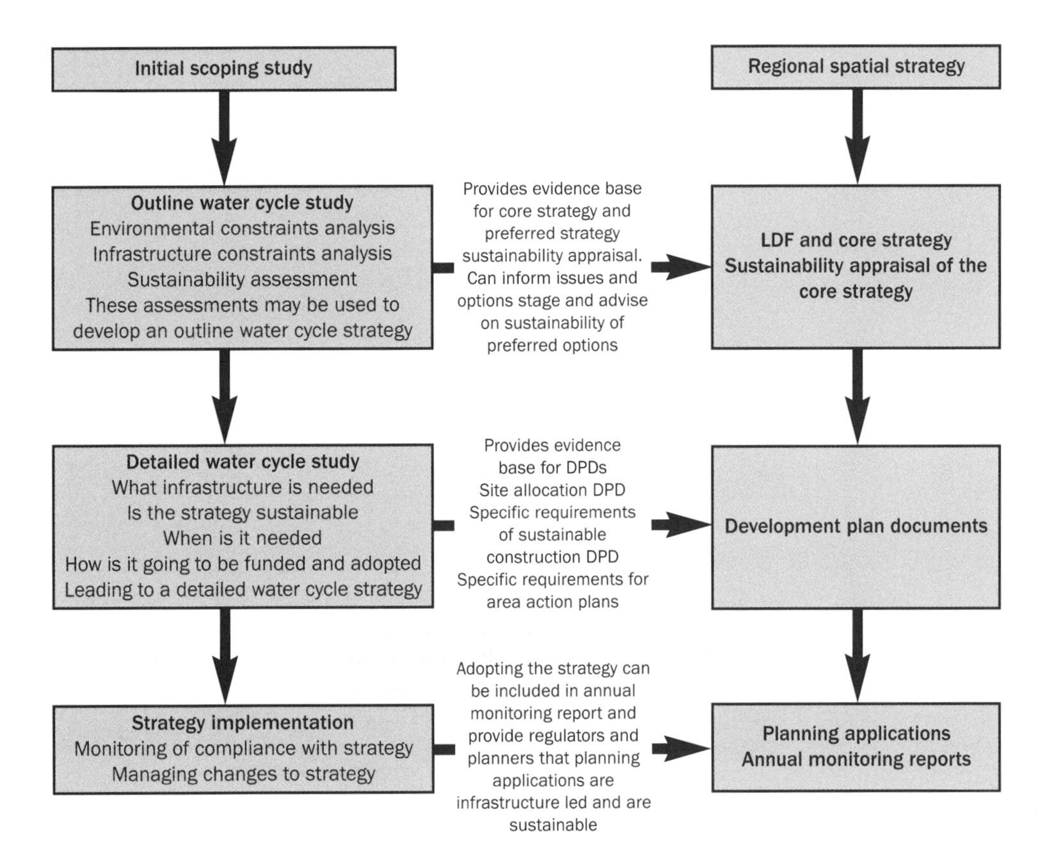

Figure A3.1 *Different stages of a water cycle study*

Most of the data and information used in a water cycle study will already exist within the organisations responsible for operating, regulating and managing the water environment. This guidance document supports the water cycle studies framework, through knowledge, information and tools relevant to different stages of a water cycle study. The WCS help planning authorities make strategic decisions about where houses should be built, what management options should be used etc. This guidance sets out the tools to help make these decisions on a strategic and site-specific basis about what the most sustainable solutions actually are, and provides the evidence base for a planning authority to support them, through the water cycle study.

Figure A3.2 provides a "navigational guide" through this publication, and highlights how it can be used in conjunction with water cycle study guidance.

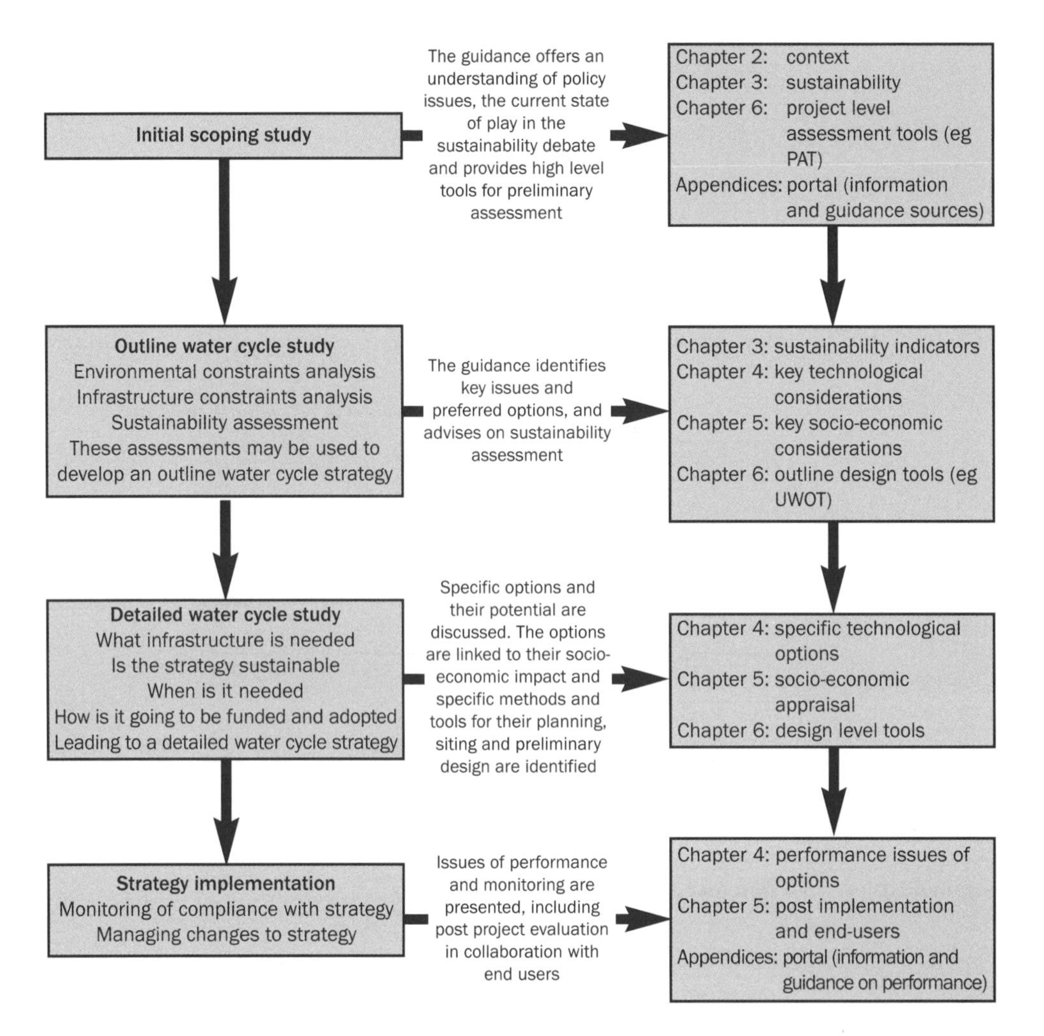

Figure A3.2 *How this guidance relates to the WCS guidance*

Links to other plans

Water cycle studies integrate separate pieces of work by different organisations, to help the planning process. It is important that the water cycle scoping study identifies what other plans and strategies have already been carried out and provides a clear route map as to how these studies integrate with, or inform the WCS.

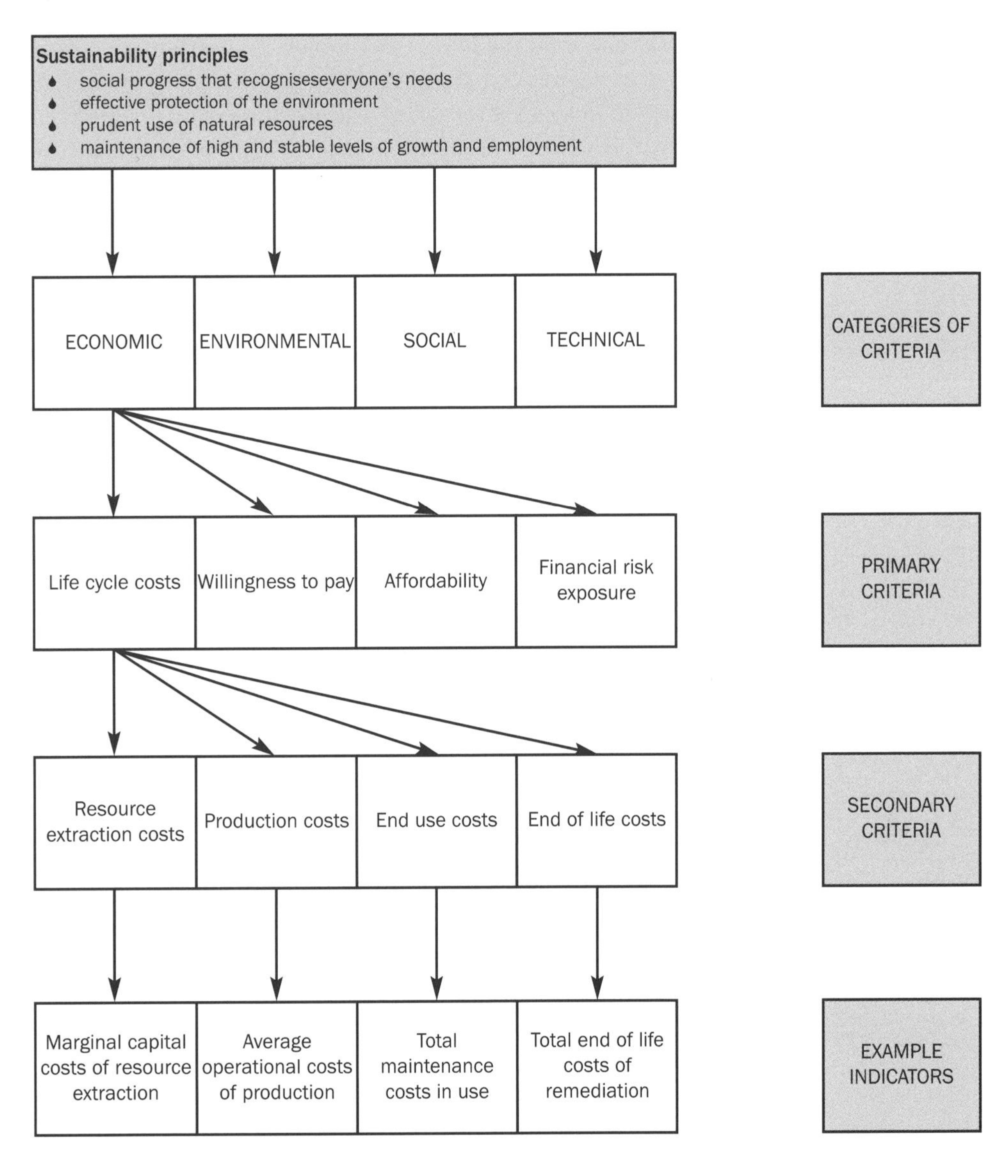

Figure A3.3 ***The relationship between sustainability principles, criteria and indicators (adapted from Ashley et al, 2004 and Foxon et al, 2002)***

Primary criteria, linked to specific sustainability principles, provide a framework to include sustainability in decision making that has been broadly accepted as a method (for example, in the regional sustainable development frameworks now incorporated into the spatial planning process).

A3.2 SCENARIOS FOR WATER CYCLE MANAGEMENT

Developing scenarios involves several assumptions, four of which were summarised in the Foresight future flooding project (Evans *et al*, 2004):

1. The future is unlike the past and is shaped by human action and choice.
2. The future cannot be foreseen but its exploration can inform decisions.
3. There are many possible futures and scenarios can map a "possibility" space.
4. Developing scenarios involves rational analysis and subjective judgement.

Foresight future was not explicitly developed for water and although prepared primarily for use by, or for the benefit of, the UK Government can be thought of as a generic example of a scenario building framework. The scenarios identify two initiatives for change of the basis for the future: social values (x axis) and systems of governance (y axis) highlighted in Figure A3.4. Social values range from individualistic values to more community orientated values, taking account of social and political priorities and patterns of economic activity that results from them. Systems of governance deal with the structure of government and the decision making process. It ranges from autonomy where power remains at a national level to interdependence where power increasingly moves to other institutions, for example, up towards the EU or down towards regional government. The two axes create four scenarios, whose names reflect their position in the space defined by the axes: world markets (interdependent and individualistic), global sustainability (interdependent and communitarian), national enterprise (autonomous and individualistic) and local stewardship (autonomous and communitarian).

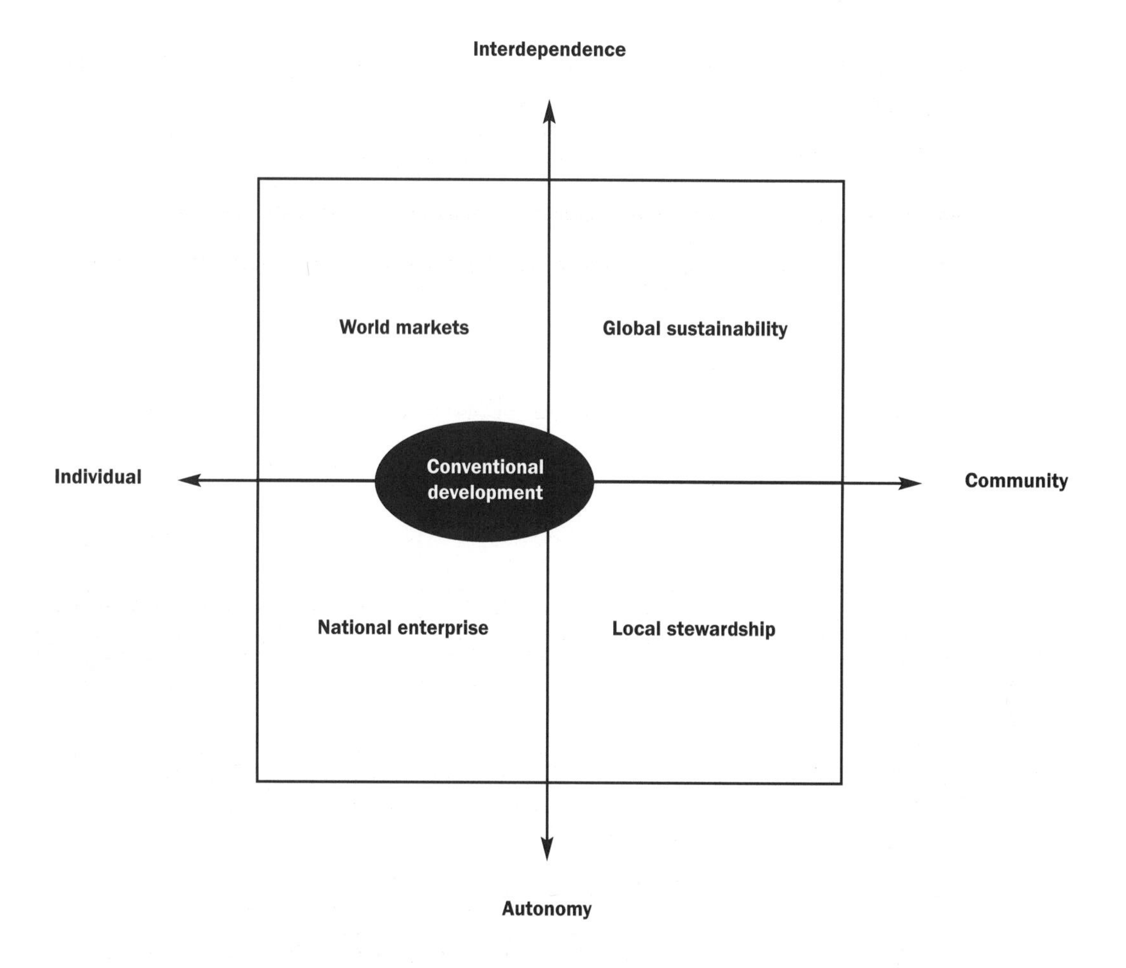

Figure A3.4 *The Foresight Future*

The scenarios have been used as a basis for strategic thinking in the UK including, for example, work on flooding (Evans *et al*, 2004), water demand management (EA, 2001b) and climate change (Holman, 2002). Significant work on the subject of future scenarios development has also been done by the Global Scenario Group (GSG), which has developed a range of global scenarios, based on a two tier hierarchy introduced in branch points: global scenarios and human choice (Gallopín *et al*, 1997 and Raskin *et al*, 2002). A combination of the Foresight futures scenarios and the GSG scenarios was used as a basis for scenario work in WaND.

A4 WaND research on water management technologies

A4.1 TECHNOLOGIES AVAILABLE FOR SWCM

Table A4.1a *Technologies available for water supply management*

Technology	Description
Water supply reservoirs	Large-scale water storage in upland locations of generally good quality water, or smaller storage in lowland locations of poorer quality river water
Groundwater abstraction	Large-scale exploitation of aquifer water of generally good quality water
Surface water abstraction (indirect wastewater reuse)	Large-scale use of (typically) river water of variable quality. Water abstracted may include treated sewage effluent form upstream urban areas
Transfer of resources	Large-scale engineered transfer of water resource from locations with abundant supply
Water-saving devices	In-house new or retrofit devices which use less water for similar function
Local abstraction	Similar to groundwater abstraction except practised on a local scale and typically for non-potable applications
Desalination	Treatment of seawater or brackish water to potable standard. High energy costs and residual brine
Dual-supply systems	Two pipe systems, one conveying potable water and one non-potable. Danger of cross-connections
Direct wastewater reuse	Treatment of sewage to potable water standards is technically possible but is extremely rare even in the most water-stressed areas
Water containers (bottled water)	The consumption of water delivered by bottle is growing at a rapid rate. Potable water delivery using bottled water is very user acceptable. Water can also be delivered in tankers, but currently only used in emergency situations in developed countries
Fit-for-purpose supply	Careful assessment of the water quality standard needs for a particular use, matched to type of water available on the site
Point-of-use treatment systems	A variety of in-house equipment (eg filters, UV disinfection, softening) for water treatment

Key	
Conventional	
Potentially sustainable and in practice	
Novel	

Table A4.1b *Technologies available for stormwater management*

Technology	Description
Combined sewers	System of sewers conveying stormwater (surface water runoff) and wastewater in the same pipe. Potential pollution form combined sewer overflows during storm events
Separate storm sewers	Two-pipe system conveying runoff and sewage in separate pipes. Danger of misconnections
Underground storage systems	Below ground storage of stormwater in tanks or oversized pipes, for controlled release after the storm event
Combined sewer overflows (CSOs)	Overflow devices on combined sewers that operate under high flows (rainfall-induced) discharging untreated but dilute sewage to watercourses
Surface detention systems	Above-ground storage of stormwater in pond, lakes etc for controlled release after the storm event
Gully pots/inserts	Sump under road gully inlets to trap sediment. Inserts designed to improve trapping performance
Wetlands	Naturally occurring vegetated waterbodies that may improve stormwater quality
Inlet control	Small-scale devices (eg disconnected downpipes, rainwater butts, surface ponding) designed to delay stormwater runoff
Swales and filter strips	Flat strips or gently side-sloped grassed ditches for stormwater conveyance and treatment
Pervious surfaces	Engineered hard surfaces allowing infiltration of stormwater in to the subsurface
Soakaways	Underground chamber filled with crushed stone allowing infiltration of stormwater into the surrounding sub-soil
Filter drains	A type of filtration trench including a buried perforated pipe for drainage popular for rural road drainage
Ponds	A type of above-ground stormwater storage area that maybe be natural or engineered for flow attenuation and quality improvement
Constructed wetlands	Engineered version of natural wetlands. Large areas required but have water quality improvement properties and aesthetic and biodiversity benefits
Sand filters	Engineered filters with sand media for stormwater quality improvement
Vegetated spaces	Areas of land set aside form stormwater collection and treatment with similar benefits to constructed wetlands
Bio-retention basins	Shallow basins used to slow and treat on-site stormwater
Sediment basins	Basin constructed to trap sediment by settling under gravity
Modular systems	Various 'off-the-shelf' treatment units such as sediment basins and constructed wetlands for stormwater treatment
Built-in storage	Proprietary devices built into the housing structure to detain stormwater, typically underground
Green roofs	Vegetated building roofs used to capture, store, release and evaporate rainfall
Evaporative SUDS	Devices designed to maximise evaporation of stormwater and minimise site runoff

Key	
Conventional	
Potentially sustainable and in practice	
Novel	

Table A4.1c *Technologies available for wastewater management*

Technology	Description
Combined sewer systems	System of sewers conveying stormwater (surface water runoff) and wastewater in the same pipe. Potential pollution form CSOs
Separate foul sewer	Two-pipe system conveying runoff and sewage in separate pipes. Danger of misconnections
End-of-pipe wastewater treatment plant	Treatment of wastewater from centralised sewer system using physical, chemical and biological processes before discharge to the environment
Cesspools	Small-scale solution consisting of closed tank for storage of wastewater, requiring regular emptying
Septic tank systems	Small-scale solution consisting of tank for storage of wastewater, followed by drainage field for dispersing flow to the subsoil. Requires less frequent emptying
Package treatment plants	A range of proprietary devices built at small scale but mimicking the processes found in end-of-pipe plants
Membrane bioreactors	Treatment process based on biological degradation in conjunction with ultra-filtration producing an extremely high quality treated effluent
Mound systems	Specialised drainage system used when subsoil properties are not ideal for septic tank installation
Constructed wetlands	Same as mentioned above (stormwater) but used for treating wastewater before discharge to watercourses
Sand filters	Same as mentioned above (stormwater) but used for treating wastewater before discharge to watercourses
Living machines	Series of biological treatment processes based on emergent vegetation and constructed within a greenhouse environment. Possible amenity benefits
Small diameter gravity systems	Small-sized foul sewers used to convey low flows or standard flows with initial solids removal (eg via a septic tank)
Low pressure sewers	Pumped outlet form toilet or septic tank conveyed under low pressure
Vacuum toilets/sewers	Vacuum pressure used as motive force draining toilets directly or via wastewater sumps
Real time control	Active control of sewers or treatment plants based on the real-time conditions, aimed at optimising performance
Air-displacement toilets	Very low water use toilets using displaced air as the main motive force
In-sewer treatment	Measures to promote and facilitate biological degradation of wastewater within the sewer system itself

Key	
Conventional	(white)
Potentially sustainable and in practice	(light grey)
Novel	(dark grey)

Table A4.1d *Technologies available for water recycling/reuse*

Technology	Description
Aquifer storage and recovery	Recharge of underground aquifers with stormwater or treated wastewater either through pumping or gravity feed, for later use
Effluent dual reticulation	Similar to dual-water supply except treated wastewater is used for the secondary supply
Rainwater harvesting	Collection and storage of rainfall from roofs and reuse for non-potable applications
Grey water systems	Collection and storage of household washwater for treatment and reuse for non-potable applications
Green roofs	Vegetated building roofs used to capture, store, release and evaporate rainfall
Combined rainwater and greywater recycling	Systems incorporating elements of both rainwater harvesting and greywater recycling
Dry toilets	Toilets requiring no or minimal water use, eg pit latrines, chemical toilets, incinerating toilets
Composting toilets	Toilets which collect faeces and urine and exploit the natural composting process to produce a natural fertiliser
Urine separation	Specially designed toilets allowing urine to be collected separately from faeces for use as a natural liquid fertiliser
Sewer mining	Pumped wastewater from local sewers for reuse applications after treatment
Autonomous housing	Houses are (almost) completely self-sufficient including in their use of water
Closed water systems	A system where all domestic wastewater is collected and treated for potable use. No known systems for practical everyday use
Energy-water systems	A range of technologies or practices that harness energy from the various water flows and in so doing achieve energy efficiency and/or treatment measures

Key	
Conventional	
Potentially sustainable and in practice	
Novel	

A4.2 GUIDANCE REPORTS ON SUITABLE WATER MANAGEMENT TECHNOLOGIES

Box A4.1 *Some guidance reports on sustainable water management technologies*

Water supply		Water efficiency	
CLG (2009)	*Code for Sustainable Homes – technical guide*	GG26R (2005)	*Saving money through waste minimisation: reduce water use*
EA (2007b)	*Conserving water in buildings: a practical guide*	Waterwise (2009)	*The water and energy implications of bathing and showering behaviours and technologies*
UKWIR Report 07/WM/08/35	*Managing seasonal variations in leakage*	Waterwise (2008)	*Evidence base for large scale water efficiency in homes*
UKWIR Report 05/WM/08/32	*Towards best practice for the assessment of supply pipe leakage*	03/WR/25/1	*Quantification of the savings, costs and benefits of water efficiency*
MTP (2007)	*Market Transformation Programme. Product overview: water*		
Stormwater management		**Recycling/reuse**	
Pratt *et al* (2002)	CIRIA C582 *Resource control using constructed pervious surfaces. Hydraulic, structural and water quality performance issues*	BS 8515:2009	*Rainwater harvesting systems. Code of practice*
Wilson *et al* (2004)	CIRIA C609B *Source control using constructed pervious surfaces. Hydraulic, structural and water quality advice*	Leggett *et al* (2001)	CIRIA C539 *Rainwater and greywater reuse in buildings. Best practice guidance*
Shaffer *et al* (2004)	CIRIA C625 *Model agreements for sustainable water management systems. Model agreements for SUDS*	Shaffer *et al* (2004)	CIRIA C626 *Model agreements for sustainable water management systems. Model agreements for rainwater and greywater reuse systems*
Woods-Ballard (2007)	CIRIA C697 *The SUDS manual*	EA (2008a)	*Harvesting rainwater for domestic uses: an information guide*
Wilson (2008)	CIRIA C680 *Structural design of modular geocellular drainage tanks*	EA (2008b)	*Greywater: an information guide*
		Leggett *et al* (2001)	CIRIA PR080 *Rainwater and greywater use in buildings. Decision-making for water conservation*
		Rachwal and Holt (2008)	FR/G0006 *Urban rainwater harvesting and water reuse: a review of potential benefits and current UK best practices*
Wastewater collection		**Policy**	
UKWIR Report 06/WM/07/6	*21st century sewerage design*	Samuels *et al* (2006)	CIRIA C630 *Sustainable water management in land-use planning*
Green and Wilson (2004)	FHRC-04 *Assessing the benefits of reducing the risk of flooding from sewers*		

A4.3 WATER SAVING DEVICES

Table A4.2 *Water consumption using conventional appliances (Grant, 2002)*

	Uses/appliances	Assumed vol/use, (litres)	How often appliance is used uses/person/day *	Measured **use litres/prop/day (l/p/d)	%
Bathroom	WC	10	4.12	52.5	35
	Bath	80	0.34	22.5	15
	Wash hand basin	6	2	12	8
	Shower	15	0.6	7.5	5
Kitchen	Kitchen sink	10	2.25	22.5	15
	Wash m/c	100	0.157	18	12
	Dishwasher	28	0.214	6	4
Other	Outside	9	0	9	6
		Total	150 litres	100%	

* Frequency values interpreted from the Environment Agency (2001a)

** % from Anglian Water survey of domestic consumption (SODCON) data 1994. lpd calculated for four person house.

Table A4.3 *Estimate of potential water saving with best available technologies not entailing excessive costs (BATNEEC) and estimated future technologies (Grant, 2002)*

	Use	Appliance frequency uses/person /day	BATNEEC (2001) vol/use	BATNEEC (2001) vol/day (l/p/d)	BATNEEC (2006) (vol/use)	BATNEEC (2006) vol/day (lpd)
Bathroom	WC	4.12	4	16.48	3	12.36
	Bath	0.34	70	23.8	50	17
	Wash hand basin	0.6	20	12	6	3.6
	Shower	2	3	6	2.5	5
Kitchen	Kitchen sink	2.25	3	6.75	2.8	6.3
	Wash m/c	0.157	45	7.07	35	5.5
	Dishwasher	0.214	18	3.85	14	2.996
Other	Outside	0	0	0	0	0
			Total	76	Total	52.8

The analysis considered the conservation effect of each option on existing and new homes separately, to account for the number of people living in the house, water-saving features and regulations that tend to differ significantly between the two. Also, water conservation options were assessed from the point of view of UK adoption, particularly with respect to climate, national experience and practices. The uptake and success of water conservation was assessed with respect to several factors including:

- absolute and relative water reduction
- cost and ease of adoption and operation
- acceptability (social, legal, health).

The expected uptake of the technologies was estimated by applying a trend analysis based on the popularity of the device and factors that would affect its uptake (eg water meters are mandatory in all new buildings). The estimations are based on current trends and behaviours. Further details are given in Sim *et al* (2007).

Table A4.4 ***Expected demand reduction effects of water efficiency options based on their performance and likely uptake in new build and existing housing stock***

Option	Current status and uptake factors	Likely uptake by 2025		Approximate reduction in per person consumption in a home compared to 2001 standard appliance (litres/person/day)	Water demand reduction potential
		New build homes	Existing homes		
Metering	28% of households 2006–2007 (regional variation)	All	Half/most	15	Major
6 litre flush toilet	Regulatory standard since 1999	All	Half	10	Moderate
Cistern displacement device (eg Hippo)	Inexpensive and easy to install	Not Fitted	Some/half	4	Moderate
Normal flow showers	Typical in new homes though power showers are also becoming more popular	Half/most	Some/half	12–20	Moderate
Reduced flow basin taps	Very low current penetration	Few/some	Few/some	6	Small
Efficient clothes washing machines	>90% of households have a washing machine (~8 year life cycle)	Most	Most	5	Moderate
Dishwashers	Low penetration	Few/some?	Some/half	4–7	Moderate
Water butts (outdoor water use)	Penetration hard to assess	Some/half	Some/half	~2.4	Small
Water efficient gardens	A feature in future new homes?	Few/some?	Few/some ?	~4.5	Small
Greywater recycling	Relatively expensive and complicated to implement	Very few/ few?	Negligible	~7.5	Very small
Rainwater collection (indoor water use)	Relatively expensive and complicated to implement	Very few/ few?	Negligible	~7.5	Very small

Assuming 2001 frequency of use for each appliance (note: few ~ 10%, Some ~ 25%, Half ~ 50%, Most ~ 75%)

Water savings and energy

The water saving and CO_2 emissions associated with some shower types are shown in Table A4.5. Further to this the suitability, advantages and disadvantages associated with different types of taps and WCs flush-volume reduction are shown in Tables A4.6 and A4.7.

Table A4.5 *Potential water savings and CO_2 emissions through shower use (adopted from EA, 2007b)*

Shower type	Flow rate	Water used per five minute shower	Water use (as % of 70 litre bath)	CO_2 emissions (kg)
4 litre/minute	4 litre/minute	20	29	0.07-0.27
7.2 kW electric	3.5 litre/minute 30°C temp rise	17.5	25	0.34 direct electric
9.8 kW electric	4.7 litre/minute 30°C temp rise	23.5	39	0.45 direct electric
6 litre/minutes (water saver)	6 litre/minute regulated flow	30	43	0.27-0.4
9.5 litre/minutes (water saver)	9.5 litre/minute regulated flow	47.5	68	0.42-0.63
Power shower	Typically 12 + litre/ minute	60+	86+	0.53-0.8+

Table A4.6 *Suitability, advantages and disadvantages of water-saving taps*

<table>
<tr><th>Tap type</th><th>Suitability</th><th>Advantages</th><th>Disadvantages</th></tr>
<tr><td>Spray taps</td><td rowspan="3">Communal buildings where frequency of use is high and duration of use is low</td><td>60-70 % water saving</td><td>• blockage of holes due to scaling, soap, grease
• legionella risk
• not suitable for filling sinks for washing up.</td></tr>
<tr><td>Push-top taps</td><td>• easy retrofitting
• hygienic</td><td>• water loss if stuck in “on” position
• not always desirable for domestic use</td></tr>
<tr><td>Battery-operated taps</td><td>• switches off automatically
• duration of flow can be adjusted</td><td>• insufficient flow duration setting can trigger second use (ie wastage of water)
• batteries required
• not always desirable for domestic use.</td></tr>
<tr><td>Single-lever mixer</td><td>Households/ communal buildings</td><td>• flexible to give restricted low and high flow</td><td>• relatively expensive.</td></tr>
<tr><td>Infra-red taps</td><td>Hospitals/ catering</td><td>• hygienic (touch free)</td><td>• expensive
• insufficient flow duration setting can trigger second use (ie wastage of water)
• batteries maybe required
• not always desirable for domestic use.</td></tr>
<tr><td>Tapmagic inserts</td><td>Households</td><td>• cheap
• flexible give full flow
• easy to retrofit
• works as spray tap at low flow</td><td>• blockage of holes
• can be removed easily.</td></tr>
</table>

Table A4.7 *Suitability, advantages and disadvantages of WCs flush volume reduction approaches*

Toilet type	Suitability	Advantages	Disadvantages
Displacement devices	Households	• easy installation • cheap.	• water waste due to incorrect installation and/or double flushing.
Siphon toilets	Households	• virtually leak free • available as low as 4.5 litres • 6 litres allowed in UK since 2001.	• the flush rate is generally lower.
Toilets with valves	Public places where frequency of use is high	• refill duration is low (time saving).	• eventually leak • high flush rate • requires a bigger diameter waste pipe.
Dual-flush	Mainly in households	• 4/6 litres toilets allowed in the UK since 2001.	• wastage due to misuse • can leak (if valve is used).
Vacuum	Areas with no access to water supply and gravity drainage	• only 1.2 l of water required • smaller bore collect • pipe and small storage area.	• high capital and operational cost • emissions.
Incinerating toilets		• 35% water saving • small amount of ash produced.	
Composting	Areas where there is no access to foul sewer	• 35 % water saving • compost may be used as fertiliser for ornamental trees.	• regular maintenance.

A4.4 WAND RESEARCH – LOW-FLUSH TOILETS

Properly designed and correctly installed toilets with flush volumes as low as four litres can be connected to conventional drains without fear that the drains will become blocked (EA, 2007b). Extensive laboratory-based trials using a full-scale test rig carried out as part of the WaND research project found that reducing the effective flush volume affects the solids transportation efficiency in a conventional 100 mm drain. Figure A4.1 shows the influence of flush volume on solids transportation in a conventional bore (100 mm) sewer.

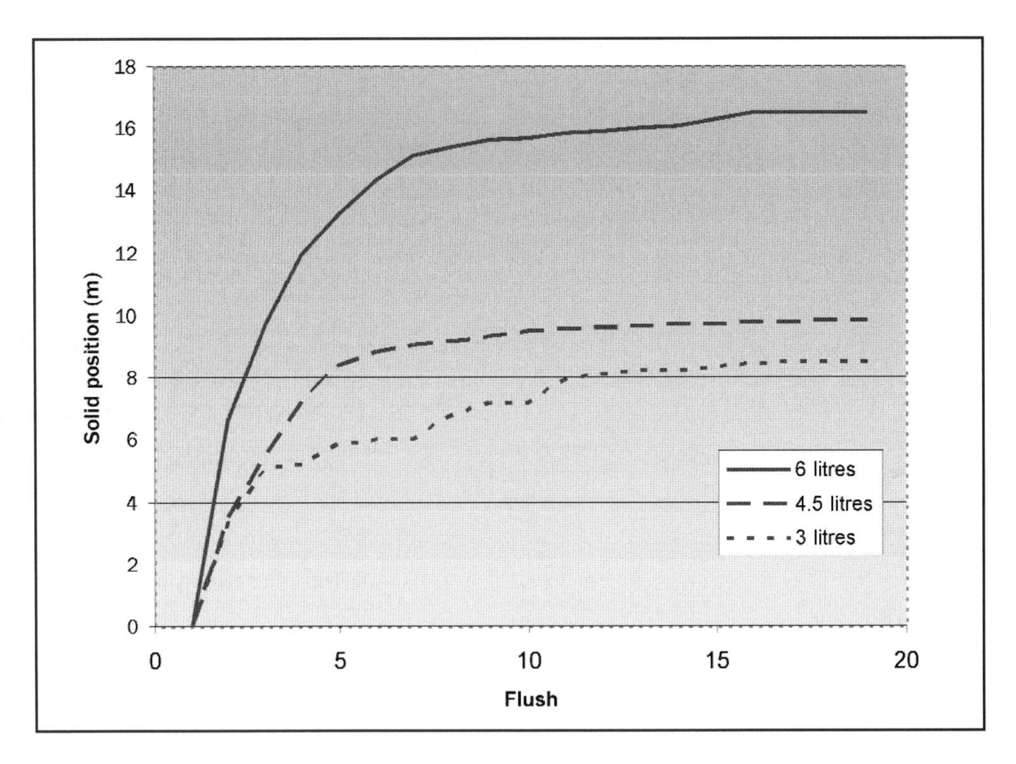

Figure A4.1 *Effect of flush volume (100 mm pipe, gradient 1:100) (source Memon et al, 2007b)*

The figure shows that the cumulative distance covered by solids subjected to low-flush volume (three litres) is less than eight metres, which is far less than the distance covered by solids when the flush volume was six litres. For flush volumes lower than four litres, care is needed with the design of the drain (EA, 2007b).

Small-bore wastewater collection system

The system comprises a prototype ultra-low-flush toilet (ULFT) and 50 mm drainage pipe. The ULFT appearance is similar to a normal toilet. It uses displaced air to flush out the waste and about 1.5 litres of water to refill the bowl seal trap. It should not be confused with compressed air or vacuum toilets that have high operational energy and extra equipment requirements. The ULFT can only be flushed when its lid is fully closed.

The toilet's flushing performance was rigorously tested at the Water Research Centre (WRc). A full-scale test rig was used to investigate a range of synthetic solids, tissue paper loadings and different pipe diameters to compare the performance with 6/4 litre dual-flush toilets. Results (Figure A4.2) showed that the toilet works exceptionally well with 50 mm diameter (ie removes the waste quickly with the least number of flushes). To assist its retrofitting, the toilet performance was also tested for 50 mm flexible pipe laid at uneven gradients and sharp bends. The performance tests showed similar results. For the conventional 100 mm pipe, the performance of ULFT can be improved by introducing a small-bore flexible pipe tail. Guidelines for ULFT installation in new and existing developments are provided in the WaND portal (Appendix A1).

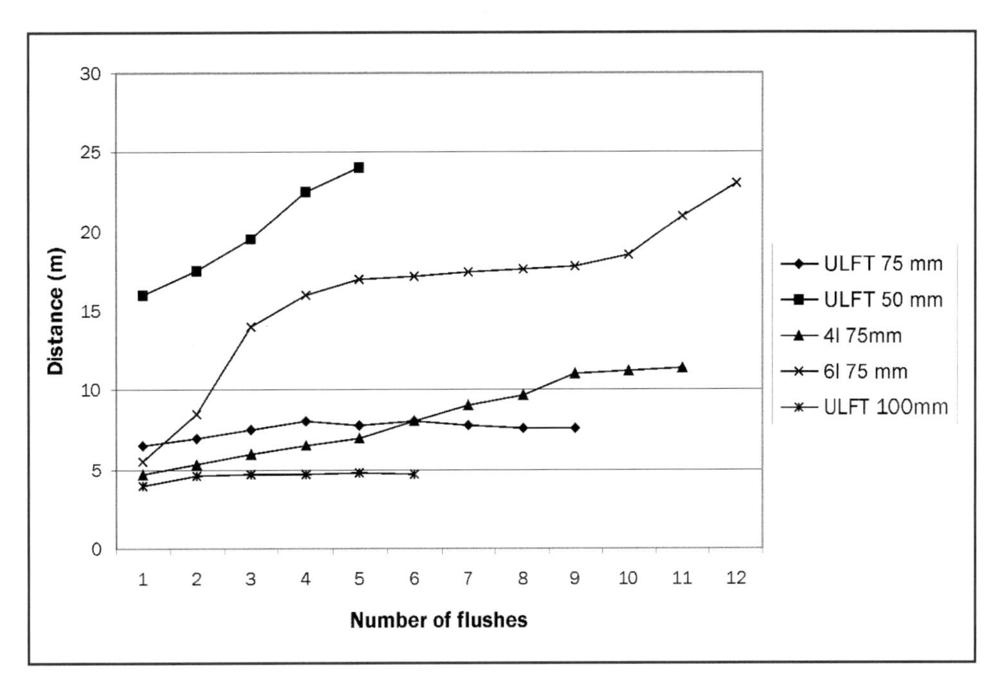

Figure A4.2 *Comparison of the performance of the ULFT and a dual-flush WC*

After establishing the system performance under controlled conditions, the system was installed in an office block (at the WRc offices in Swindon) to monitor its performance under real operating conditions. Box A4.2 is a summary of the trial and main observations made with respect to its performance both in terms of water saving and user acceptance.

Box A4.2 *Small-bore wastewater collection system field trials – WRc case study*

Given the innovative nature of the ULFT, a study was carried out to investigate public acceptance of the appliance before attempting wider-scale rollout. The study to assess users' acceptance and the water-savings potential of the ULFT took place over an eight month period at the Water Research Centre (WRc) in Swindon. During this time, a toilet block consisting of five conventional WCs (two male and three female) was monitored to record actual water consumption by WCs and their frequency of use. The study was divided into two stages: Stage 1 (two months) and Stage 2 (six months).

The objective of Stage 1 was to ascertain the two WCs that should be retrofitted with ULFT. The criteria used to choose the WCs were "popularity" (determined by the frequency of use) and water consumption (determined by the volume per flush).

Stage 2 included retrofitting ULFTs, establishing users' perception of the ULFT and continuing with the monitoring activities in the toilet block that had begun in Stage 1. User perception was analysed taking two factors into account: user preference and user acceptance. The former refers to the popularity of the ULFT when compared to a conventional WC, and was measured by comparing the proportional frequency of use of the two types of WC during Stage 1 and Stage 2 of the trials. User acceptance was assessed by means of short questionnaires designed to be answered by users of the ULFT after their visit to the lavatory.

- the questionnaires used for this exercise included a brief description of how the ULFT works and instructions for its use (closing the lid before flushing)
- the questionnaire consisted of ten multiple-choice questions and a section for comments
- the questions were targeted at evaluating issues such as whether the respondent valued water conservation, ease of operation, flushing and cleaning performance, and comfort and design of the WC
- eight months of data showed potential waste and energy coming up to 85 per cent compared to nine litre conventional WC and there were no repeated blockages
- over half of the respondents to the perception survey thought ULFT were easy to use.

However, some users viewed the need to close the toilet as unhygienic and were concerned about not being able to know when the lid was properly closed.

Environmental performance

A comparative performance of conventional WC and the ULFT used in the trials is shown in Table A4.8. The performance is expressed in terms of potential water savings and consequent energy savings and reduction in CO_2 emissions.

Table A4.8 *Water and energy savings with ULFT s (relative to conventional WC)*

<table>
<tr><th colspan="2">WC</th><th rowspan="2">Ave. water used (l)</th><th rowspan="2">Water saved (l)</th><th rowspan="2">Water saved (%)</th><th rowspan="2">Energy used (Joules)</th><th rowspan="2">Energy saved (Joules)</th><th rowspan="2">Energy saved (%)</th><th rowspan="2">CO_2 produced (g)</th><th rowspan="2">CO_2 saved (g)</th><th rowspan="2">CO_2 saved (%)</th></tr>
<tr><th>ID</th><th>Type</th></tr>
<tr><td rowspan="2">m</td><td>Conventional WC</td><td>10.3</td><td rowspan="2">8.9</td><td rowspan="2">87</td><td>33524</td><td rowspan="2">28561</td><td rowspan="2">85</td><td>4.15</td><td rowspan="2">3.54</td><td rowspan="2">85</td></tr>
<tr><td>ULFT</td><td>1.37</td><td>4963</td><td>0.61</td></tr>
<tr><td rowspan="2">f</td><td>Conventional WC</td><td>10.0</td><td rowspan="2">8.7</td><td rowspan="2">87</td><td>32710</td><td rowspan="2">27975</td><td rowspan="2">86</td><td>4.05</td><td rowspan="2">3.47</td><td rowspan="2">86</td></tr>
<tr><td>ULFT</td><td>1.3</td><td>4735</td><td>0.59</td></tr>
</table>

Note: m and f represents ULFT fitted in male and female cubicles

A4.5 RAINWATER HARVESTING

Box A4.3 *Examples of RWH systems installed in the UK*

Brewer *et al* (2001) identified over 20 RWH systems with an objective of monitoring their performance to investigate practical issues and develop cost-benefit analysis. Since publication of these case studies the technology has moved on, but the case studies and presentation of cost-benefit analysis presented in Table A4.11 is indicative of the main considerations.

Example 1: Office building: the office building was situated in a rural location in Bedfordshire. It had about 50 occupants in 1500 m² of two storey office accommodation.

The building was designed to reflect the organisation's commitment to environmental best practice and resource conservation. Energy and water efficiency measures were incorporated into the design from conception. The primary motivation was a desire to comply with general environmental policies, rather than to provide a demonstration of good practice in construction.

The RWH system comprised of total catchment roof area of 800 m², an 8000 litre underground collection tank, a treatment unit (filtration and UV disinfection), a 750 litre cistern for treated water, pumps and a network of pipes and gutters. Other water conservation measures, in addition to RWH, include low-flush toilets and urinals. Rainwater was supplied to 12 WCs and four urinals.

Example 2: Ecological housing development: the development is located near Newark-on-Trent, Nottinghamshire and consists of five terraced houses. The aim of the development was to create homes based on sustainability principles using minimal energy and with little environmental effect. Each of the houses is fitted with dual-flush toilets, aerated taps and efficient washing machines.

The site is self-sufficient in its water needs, all being supplied by rainwater.

Table A4.9 *RWH system maintenance activities and corresponding frequency (Leggett et al, 2001)*

Maintenance	Frequency
Manual cleaning filters	Monthly
Self-cleaning/coarse filters	3 months (check and clean)
Roof and gutter	1–2 times/year
Cartridge filters	Replace after 3 months
UV lamp	6/12 months replacement
Chemical disinfection	Replace after 3 months
Pump	Annual check wiring and function

Cost-benefit analysis for RWH

Table A4.10 provides a summary of cost-benefit for the two systems described in Box A4.3. The table indicates that for the systems investigated, the payback period is very long. This was a result of the systems being designed conservatively with very large collection tanks. Improved payback periods are now possible, particularly for efficiently designed commercial systems where demand for non-potable use is high (Table A4.11).

Table A4.10 *Cost-benefit analysis of RWH systems described in Box A4.3*

	Office building	Ecological houses (per house)
Non-potable use	376 m^3 (WC)	377 m^3
Rainwater used	150 m^3	377 m^3
System capital cost	£7250	£11 854
Operating cost	£184	£110
Value of water saved	£241	£511
Annual saving	£57	£401

Table A4.11 *Examples of two rainwater harvesting schemes with short payback periods (EA, 2008a)*

	Office in Manchester	Community Centre in Kent
Roof area (m^2)	3200	950
Rainfall (mm)	806	728
Tank size (m^3)	110	26
Amount of water collected (m^3)	2323	510
Supplying	WC's for 550 employees	WC/clothes washing
Capital cost (£)	12 000	6500
Annual water savings (£)	4000	2200
Payback* (years)	3	3

Note: Payback calculation excludes maintenance and running costs.

A4.6 GREYWATER

WaND research – substances discharge into greywater

The study found that overall the only domestic products seen to be a potential risk to reuse treatment technology are bleach and car oil with secondary risks associated with washing powder and vegetable oil (Knops *et al*, 2007). Table A4.12 gives information on the critical concentrations (concentrations beyond which the bacterial activity slows down considerably) of substances that could appear in greywater. The information can be used to educate users of greywater systems on products that should be disposed of separately or used in a way that ensures they do not enter the greywater.

Table A4.12 *Critical concentrations of pollutants affecting microbial activity*

Substance	ml/l	Substance	ml/l
Bleach	1.4	Carpet cleaner	30
Caustic soda	4.5	Pet shampoo	36
Vegetable oil	<10	Bathroom cleaner	43
Perfume	12	Food (soup)	60
Car oil	15	White spirit	95
Washing powder	24	Alcohol (spirit)	130
Hair dye	<26	Make up remover	320

The WaND research project investigated the suitability of five technologies for the treatment of greywater (Boxes A4.4 to A4.7). These were:

- a membrane bioreactor (MBR)
- a membrane chemical reactor (MCR) based on an advanced oxidation process (TiO2/UV)
- horizontal-flow reed bed (HFRB)
- vertical-flow reed bed (VFRB)
- "GROW" green roof water recycling system.

Box A4.4 *Pilot scale MBR system – construction and operating conditions*

The MBR consisted of two joint Perspex reactors of 34 litres each. The biomass seeded in the reactors came from an activated sludge system located in the pilot hall at Cranfield University. Each reactor was fitted with two submerged A4 flat sheet Kubota membranes. Aeration systems were fitted under the two pairs of membranes to provide scouring and limit the build up of a cake layer on their surface and to provide air to the bacteria. A recirculation loop generated by air lift was integrated on the front of each reactor to provide mixing of the biomass. The air flow rates of the aeration and air lift were set at 5 and 10 l/min respectively. For the whole duration of the trials, the MBR was run at a flux of 15 l/m^2 per hour, which corresponds to a hydraulic retention time of 9.7 hours. Biomass was wasted corresponding to a solids-retention time of 68 days and the biomass concentration (MLSS) was then stabilised at 8817±733 mg/l.

Box A4.5 *Pilot scale MCR system – construction and operating conditions*

The MCR was composed of a 9 litre stainless steel reactor in which a dose of 5 g/l of titanium dioxide Hombikat UV-100 (Sachtlebaen Chemie GmbH) was added to the water to be treated. Mixing of the slurry was provided by an aeration system at the bottom of the reactor. For this the air flow rate was set at 5 l/min. For the photo-catalytic reaction, four 25W UVC lamps (UVO3, UK) previously cased in glass tubes were dipped into the reactor. The slurry was then re-circulated into a loop containing a tubular membrane (inside/out system, 10 lumens of 5 mm diameter, 0.05 μm pore size and 0.157 m² surface area). The performance was investigated for a flux of 15 l/m²h which corresponds to a HRT of 3.8 hours. The MCR schematic is shown in Figure A4.3.

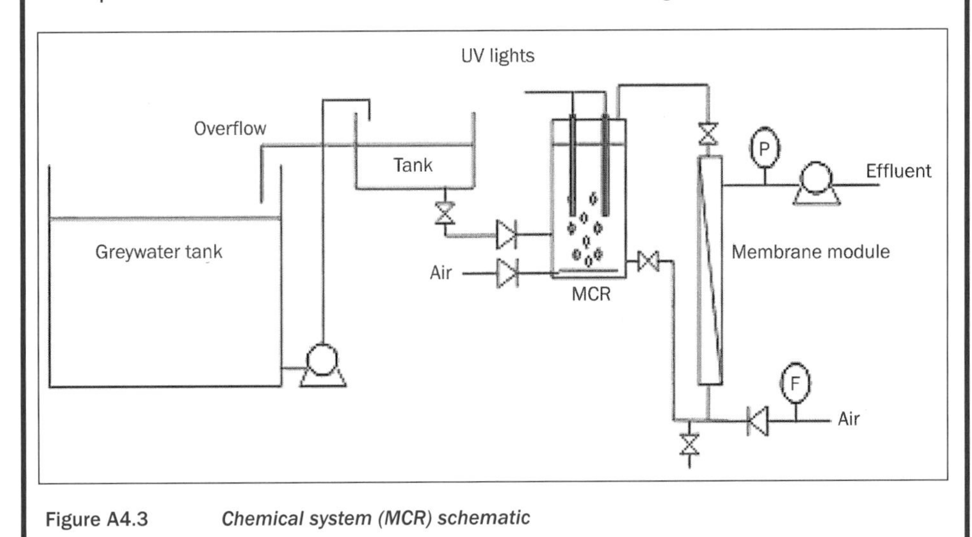

Figure A4.3 *Chemical system (MCR) schematic*

Box A4.6 *Pilot scale VFRB and HFRB system –construction and operating conditions*

Both reed beds were planted with Phragmites australis in a sand:soil:compost medium (ratio 65:25:10) with coarse (20 mm) gravel at the inlet of the HFRB and at the outlet zone of both systems. The two systems were otherwise identical with a surface area of 6 m² and a media depth of 0.7 m. The systems were fed with 480 l/d of greywater, continuously for the HFRB and supplied as 10 batches over 24 hours for the VFRB corresponding to retention times of 2.1 days and 2 hours for the HFRB and the VFRB respectively.

Figure A4.4 *Pilot scale reed bed*

Box A4.7 *Pilot scale GROW system-construction and operating conditions*

> The GROW system (WWUK, 2003) to cleanse grey water to 'green' water has been designed to sit on a flat or pitched roof suitable for use in urban environments where land space is limited and costly. GROW is a hybrid between a vertical and horizontal flow reed bed. The pilot rig comprised of five rows of two troughs in series. Troughs were filled with a 10 cm deep layer of Optiroc (lightweight, expanded clay) and topped with about 6 cm of gravel chippings (10 mm – 20 mm diameter) and planted with different types of aquatic or marginal plants (Table A4.13) as shown in Figures A4.5 and A4.6.
>
> The entire system is covered with a reinforced membrane to stop rainwater entering and to restrict light reaching the roots of the plants to prevent uncontrolled propagation. The greywater to be treated enters the inlet well at the top corner of the first trough. At the end of each row of troughs, the water flows down a weir into the lower well. A screening step is used at the inlet to remove hair and other large particles that could potentially clog the system. A coarse mesh has also been installed at the outlet to stop media being flushed out with the treated effluent. GROW is designed to treat around 1.4 m^3 per day, sufficient green water for 40–50 persons toilet flushing requirements. The green water is dyed a light tinge of green with inert vegetable colour to distinguish it in use from potable water.

Table A4.13 *Plants in the GROW system*

Trough	Plants	
	Number	Type
1	-	unplanted
2	6	Iris pseudocorus
3	8	Veronica beccabunga
4	6	Glyceria variegates
5	6	Juncus effusus
6	6	Iris versicolor
7	7	Caltha palustris
8	8	Lobelia cardinalis
9 and 10	7	Mentha aquatica

Figure A4.5 *GROW system construction*

Figure A4.6 *GROW system operation phase*

These technologies were fed from the same source of greywater and their pollutant removal efficiency was monitored at regular intervals. The influence of shock loadings was also investigated.

Table A4.14 *Compliance level of the systems for different standards and guidelines*

Location/ organisation	Parameter	Standard	Compliance level (%)		
			MBR	MCR	VFRB
WHO[1]	Faecal coliforms	1000	100	100	100
USEPA[2]	Faecal coliforms	14 for any sample (0 for 90%)	100	100	76 (36)
BSRIA[3]	Faecal coliforms	nd*	100	100	36
Japan[4]	Turbidity	2	100	93	68
	E. Coli	nd*	100	100	13
Israel[5]	BOD	10	100	54	84
	SS	10	100	100	50
	Faecal coliforms	1	100	100	39
Spain, Canary Islands[2]	BOD	10	100	54	84
	SS	3	80	79	10
	Turbidity	2	100	93	68
	Total coliforms	2.2	75	100	0
Queensland, Australia[6]	BOD	20	100	92	96
	SS	30	100	100	100
	Total coliforms	100	100	100	0

Notes
1 Hespanhol and Prost, 1994
2 USEPA, 2004
3 Mustow and Grey, 1997
4 Tajima, 2005
5 Gross *et al*, 2007
6 Queensland, 2003
* Not detectable.

Environmental implications of greywater systems

An environmental performance evaluation using a standard life cycle analysis tool was employed to investigate the effect of construction and operation of the four technologies. The results (Table A4.15) show that energy intensive technologies have higher environmental footprint than greener technologies.

Table A4.15 *The characterised effect* (development scale: 500 households)*

Impact category	Unit	GROW	Reed beds	MBR	MCR
Abiotic depletion	kg Sb eq	381	533	734	974
Global warming	kg CO_2 eq	48 100	68 600	96 100	149 000
Ozone layer depletion	kg CFC-11 eq	0.0106	0.0684	0.0187	0.0263
Human toxicity	kg 1,4-DB eq	9460	13200	23700	37 400
Fresh water aquatic ecotox	kg 1,4-DB eq	823	1510	2080	6940
Marine aquatic ecotoxicity	tons 1,4-DB eq	19 900	16 500	66300	87 900
Terrestrial ecotoxicity	kg 1,4-DB eq	107	135	315	462
Photochemical oxidation	kg C_2H_2	17.9	15.2	26.9	67.6
Acidification	kg SO_2 eq	487	387	731	1710
Eutrophication	kg PO_4— eq	31.5	40.5	31.7	60.2

Note: * the effect is shown as equivalent of specific emissions of substances described in the second column.

Costs of greywater systems

The cost of a technology is as important as its treatment performance for development of full-scale applications. Although it was not always possible to determine the full capital cost of the technologies investigated, it is possible to estimate the operational costs (cost of energy and consumables). The estimated energy consumption for different system components for the investigated technologies is shown in Table A4.16.

Table A4.16 *Energy consumption of the investigated technologies*

Technology component	Daily energy consumption (kW.h)				
	MBR	MCR	VFRB	HFRB	GROW
Centrifugal pump	6.00	6.00	0.19	–	–
Peristaltic pump	2.40	2.40	–	2.40	2.40
Air pump	0.96	0.96	–	–	–
UV lamps	–	0.60	–	–	–
Computer + data logging	12	12	–	–	–
Total	21.36	21.96	0.19	2.40	2.40
Cost (pence per day)*	181.6	186.7	1.6	20.4	20.4

Note: = not applicable, * = based on 8.5p per kW.h (DTI, 2006).

Consumables were only required for MCR. In the MCR, powdered titanium dioxide was used for the photo-catalytic reaction at a dose of 5 g/l. Considering a price of £17.5 per kg (Li Puma *et al*, 2007), the cost of the titanium dioxide needed for the 9 l reactor was £0.79. In the system, the titanium dioxide is regenerated by the UV lamps and is prevented from being washed out of the reactor by the membrane so it is reused.

Box A4.8 *Examples of greywater recycling systems used for economic assessment*

Example 1 – small Scale GWS: the system was installed in a new-build five bedroom house near Maidenhead, occupied by three adults, three children under 15 years old, and three dogs. The GWS consisted of a collection tank and a cistern. Greywater from two baths, two showers and three hand basins is collected in an underground collection tank. Greywater flows to this tank under gravity and is filtered through a fine mesh screen of 26 μm aperture, before being pumped to the cistern in the loft. Jets of filtered greywater cleaned the filter whenever the pump is activated to send water to the cistern. A bromine tablet dosing system was used to disinfect the stored greywater, which feeds five WCs and is automatically topped up with mains water if necessary. Meters were installed to record consumption of treated greywater and mains water and electricity by the recycling system. Operational maintenance included the annual addition of disinfectant to the dosing system in the roof. The self-cleaning filter also required an annual check to remove any debris (hair etc) accumulated on its surface (Brewer *et al*, 2001).

Example 2 – large scale GWS: the second system was a demonstration full-scale plant installed in a hall of residence at University of Loughborough serving 40 residents. The plant had the raw greywater buffering tank with capacity of 1400 litres. There were also two treated water storage tanks, a low-level tank (700 l) attached to the treatment plant and a high-level tank (500 l) connected to WCs. Meters were installed to record consumption of treated greywater and mains water and electricity by the recycling system. The system was composed of four main stages and a fifth stage as a polishing stage. Stage one was a preliminary treatment in the form of a balancing tank and screens for the removal of large suspended solids, floating matter and grit. The second stage was primary treatment consisting of a roughing filter and an up-flow anaerobic tank. The third stage was for biological treatment and comprises a combination of aerobic suspended and attached growth processes. The final stage of tertiary treatment was a combination of physical and biological process (Surendran and Wheatley, 1998).

Table A4.17 *Total capital cost for the small and large-scale example systems (Memon et al, 2005)*

	Small scale system (£)	Large scale system (£)
Site preparation cost	75	905
Purchase of components	1000	645
Collection and distribution pipework	150	385
Installation and commissioning	400	1410
Estimated total initial capital cost	1625	3345

Table A4.18 *Operation and maintenance cost and savings (Memon et al, 2005)*

	Small scale system (£)	Large scale system (£)
Volume of water saved (m^3)	31	420
Annual cost of consumables (£)	20	Included in labour cost
Inspection and maintenance cost (£)	60	85
Electricity consumption (kWh/year)	58.9	774
Electricity charges (pence/kWh)	6.79	5.58
Annual electricity operating cost (£)	4	43.2
Mains water charges (£/m^3)	0.69	0.81
Sewage disposal charges (£/m^3)	0.42	0.42
Unit O&M cost (£/m^3)	((20 + 60 + 4)/31) = 2.71	((85 + 43.2)/420) = 0.3
Value of water saved (£/year)	31 × (0.42 + 0.69) = 34.4	420 × (0.42 + 0.81) = 516.6

The tables highlight that the GWS installed at a larger scale are more economically viable. They may incur high installation costs but they have greater payback benefits associated with the water they save. Detailed guidance for greywater recycling systems design, installation, operation and maintenance is readily available from different sources (Box A4.9).

A4.7 STORMWATER MANAGEMENT

Box A4.9 *SUDS documents*

ELLIS, J B, SHUTES, R B E and REVITT, M D (2003a) *Constructed wetlands and links with sustainable drainage systems*, R&D Report P2-159/TR1, Environment Agency, Bristol

ELLIS, J B, SHUTES, R B E and REVITT, M D (2003b) *Guidance manual for constructed wetlands*, R&D Report P2-159/TR2, Environment Agency, Bristol

HATT, B, DELETIC, A and FLETCHER, T (2004) *Integrated stormwater treatment and reuse systems*, Technical Report 04/1, Inventory of Australian practice. Available from: <http://iswr.eng.monash.edu.au/research/projects/stormwater/hattwsud04.pdf>

KELLAGHER, R B and LAUCHLAN, C S (2005) *Use of SUDS in high density developments*, Report SR 666, version 3.0, HR Wallingford

MELBOURNE WATER (2005) *Water sensitive urban design-engineering procedures: stormwater*, CSIRO Publishing, Australia (ISBN: 978-0-64309-092-7)

WOODS-BALLARD, B (2007) *The SUDS manual*, C697, CIRIA, London (ISBN: 978-0-86017-697-8)

WOODS-BALLARD B, KELLAGHER R *et al* (2007) *Site handbook for the construction of SUDS*, C698, CIRIA, London (ISBN: 978-0-86017-698-5)

Stormwater management – environmental impact of SUDS

Figure A4.7 and A4.8 highlight the environmental impacts of SUDS based on an outline assessment for a range of SUDS performance compared to conventional drainage systems (Voaden, 2008). Figure A4.7 suggests that relatively filter drains have the poorest environmental performance. However, Figure A4.8 suggests that SUDS have significantly lower environmental impact compared to conventional drainage scheme

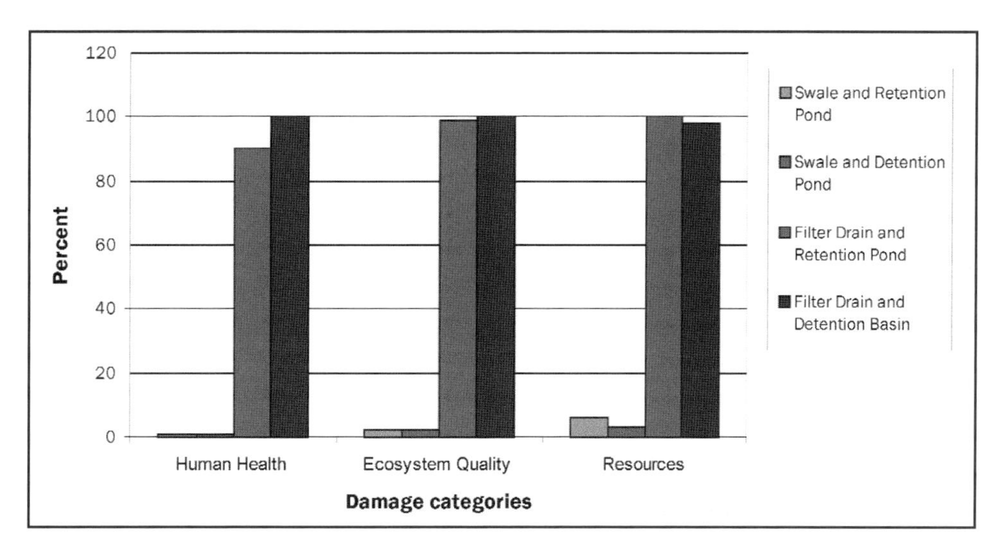

Figure A4.7 *Environmental effect of different types of SUDS*

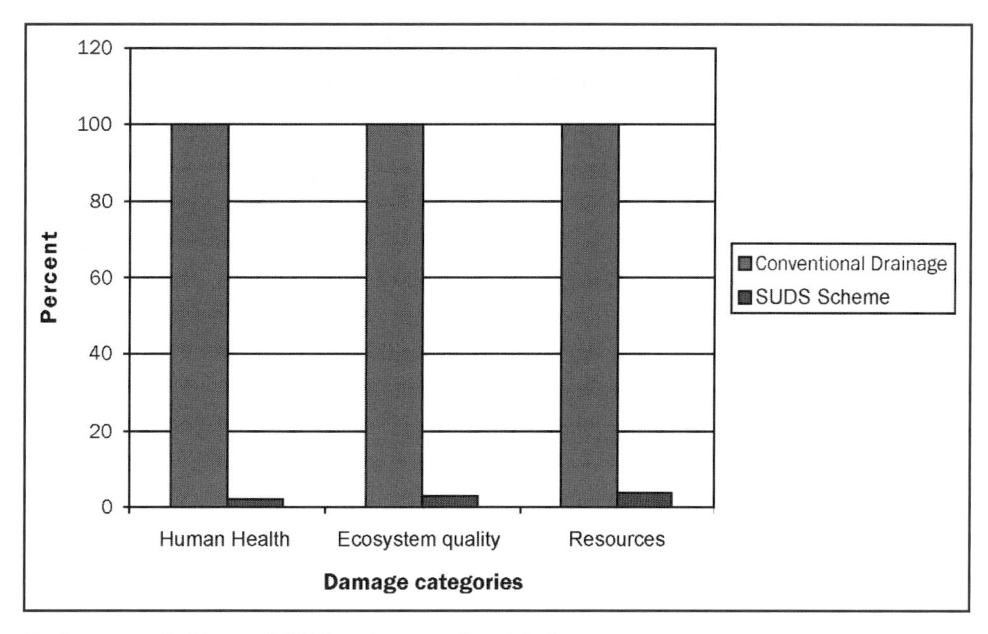

Figure A4.8 *Environmental effects of SUDS and conventional drainage*

WaND Research – rainwater storage

The WaND research project investigated the influence of rainwater storage provision, and its later use for non-potable applications, on surface runoff volume reduction. This was achieved by developing an Infoworks CS based simulation using 100 years stochastic rainfall time series (Kellagher and Maneiro Franco, 2007). The series was processed to derive a three year rainfall time series and a series of 10 extreme rainfall events (on the basis of rainfall depth collected within six hours). The simulations were run to capture the influence of catchment area (20 m^2/person), storage volume provided per person (0.75 m^3 and 1.5 m^3 per person), non-potable demand (25, 50 and 100 litres per person per day), the total area of roads (0.2 ha) contributing runoff to the storm sewer and finally seasonal variations. The simulation was run for a development with 100 inhabitants. The simulation results using long-term time series rainfall data suggest:

- the roof area per person is a critical factor in achieving reasonable guaranteed yields
- a roof area of 20 m^2 per person in drier parts of the UK can reduce the average daily water by around 25 litres per person per day. About 75 per cent of the collected

rainfall runoff is used even for the smallest demand rate, so there is less than 25 per cent of the volume going to runoff. Receiving streams of the residual capacity in a tank is important, so high use of rainwater in the property is vital to obtain maximum stormwater benefits

- the larger storage unit can provide up to 50 per cent reduction in peak flow
- although a 50 per cent reduction is significant, attenuation requirements for storm flows from a site requires significantly greater attenuation so there is a need for "downstream" systems for extreme event control
- retro-fitting urban areas with rainwater use systems, which suffer from relatively frequent flooding, could significantly reduce the flooding frequency and the flooding volumes.

The effect of extreme storm events was also assessed. Figures A4.9 and A4.10 illustrate the effect of non-potable water demand and rainwater storage allowance on reduction in runoff volume. The figures suggest that for extreme events, the reduction in runoff volume is proportional to the collection tank capacity and is rather insensitive to per capita demand for non-potable uses.

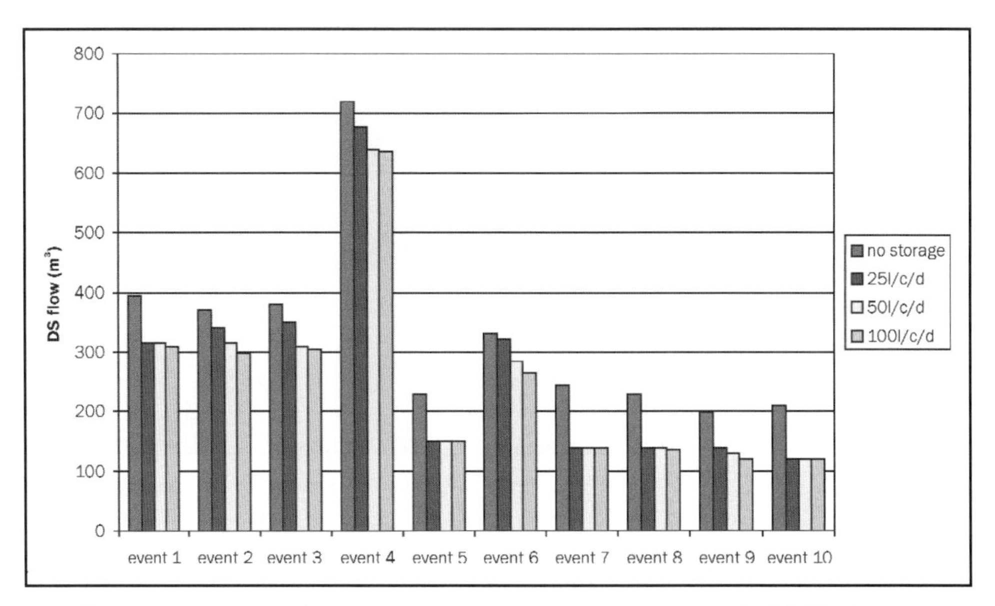

Figure A4.9 *Runoff reduction in volume from roads and buildings with a storage tank of 0.75m³/person*

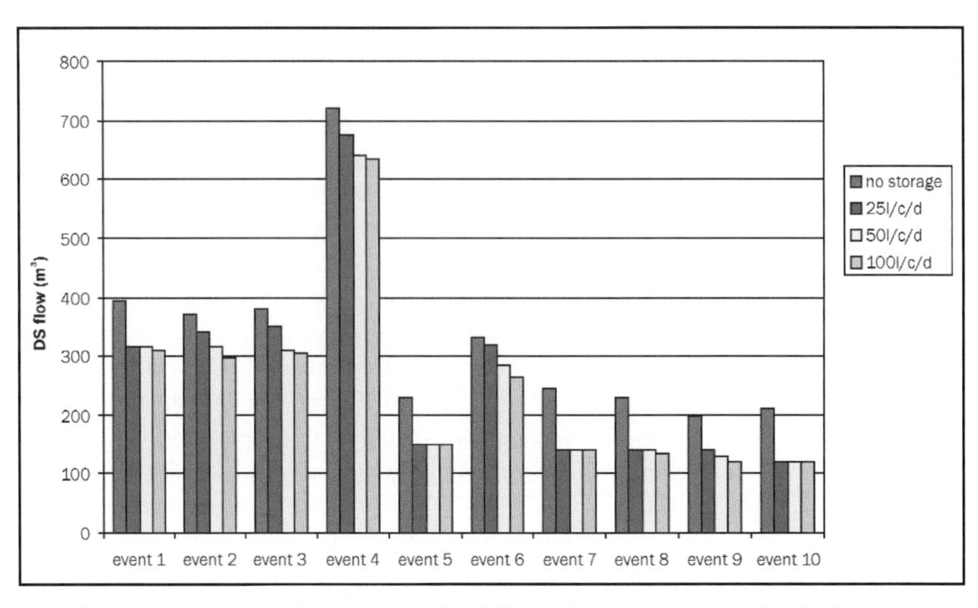

Figure A4.10 *Runoff reduction in volume from roads and buildings with a storage tank of 1.5m³/person*

A5 List of collaborators involved in the WaND research

Table A5.1 *Organisation and collaborators involved in the WaND research*

Organisation	Name
Cambridgeshire County Council	Barbara Wilcox
MWH	David Balmforth
House Builders Federation	Kendrick Jackson
Severn Trent	Paul Griffin
Environment Agency	Phil Chatfield
Yorkshire Water	Stephanie Walden
Thames Water	Sian Hills
Veolia Water/Three Valleys	Peter Greenaway
WSP Development	Alastair Atkinson
Essex and Suffolk Water/NWL	Jenny Abel
Hydro International	Mike Faram
Phoenix Product Development LTD	Garry Moore
Kirklees Metropolitan Borough Council	Andrew Jackson
Yorkshire Water	Stephanie Walden
Environment Agency	Suresh Surendran
Environment Agency	Magdalena Styles
Bucks CC	Steve Orchard
Other organisations involved : ARUP, CIRIA, Gallagher Estates, MWH, Newcastle City Council, Oceans ESU, Scottish water, SEPA, SNIFFER, Three Valleys Water, UKWIR, Wallingford Software, WaPUG, WWUK	

Table A5.2 *WaND project te*

Organisation	Researcher
Centre for Ecology and Hydrology	James Blake John Packman
Aberystwyth University Centre for Research Into Environment and Health	Lorna Fewtrell David Kay
HR Wallingford	Elena Maneiro Franco Richard Kellagher
Imperial College London	Surajate Boonya-Aroonnet
University of Bradford	Cathy Knamiller Christine Sefton Dave Luckin Heidi Smith Liz Sharp Sam Wong
University of Cranfield	Bruce Jefferson Geraldine Knops Lisa Avery Marc Pidou Nathalie Bertrand Paul Jeffrey Ronnie Frazer-Willaim
University of Exeter	Abdi Fidar Ana Maria Millan Christopher J Shirley-Smith Christos Makropoulos David Butler Dragan Savic Fayyaz Ali Memon Mark Morley Sarah Ward Shuming Liu
University of Leeds	Adrian McDonald John Parsons Patrick Sim Phil Rees Jianhui Jin
University of Sheffield	Adrian Cashman Louise Hurley Richard Ashley Steve Mounce Virginia Stovin
WRc	Carmen Waylen Kim Littlewood Nick Orman

Table A5.2 (contd) *Others involved in WaND*

Other researchers involved	Bob Andoh Caroline Bird Issy Caffoor Mike Farrimond Mark Fletcher David Fortune Ben Gersten Ian Hardwick Colin Percy Mike Pocock Sarah Reid Clare Ridgewell Paul Shaffer Suresh Surendran Rob Wesrcott Tom Wild

A6 LANDCOMs list of treatment technologies and their associated attributes

Clustered development

Single households

Water treatment technology	Typical scale[1]	Treatment technology	Other information	Water quality suitable for[2]	Footprint (m^2)	Capital expenditure ($000)	Operating costs (per year)
Nubian	0.5–1.1 kL/d (2·6 EP)	Biological filtration followed by membrane filtration	Greywater treatment		3	5 + installation	Low
Perpetual water	0.5–0.7 kL/d (2·6 EP)	Physical – sedimentation followed by adsorption	Modular greywater system		1.5	6.5 + installation	$365
Clearwater aquacell	0.5–100 kL/d (2·500 EP)	Membrane bioreactor	Modular system catering to wide range of scales		1.2–2.4	13 (single house) 100 (500 EP)	$500 (6 EP) $5500 (500 EP) $10 000 (1000 EP)
Biolytix	0.5–10 kL/d (2·50 EP)	Natural – humus filter situated at each household	Decentralised treatment	Subsurface irrigation	2.4–16	13 per house	$400
Biolytix (+LIF)	12–100 kL/d (60·500 EP)	Natural – humus filter coupled with a modular ultrafiltration unit	Combines decentralised treatment with reuse opportunituies		35–500	30 for 50 EP	$1200
Novasys – BIOSYS	1–150 kL/d (5·1000 EP)	Biological – fixed film bioreactor	Further treatment required for water to be used for non-potable water	Restricted irrigation	2	–	–
Rootzone (vertical filler – greywater	0.5–360 kL/d (2·1800 EP)	Subsurface wetland with a vertical recirculating filter	Land intensive treatment process	Disinfection required	2–500	5 (single house) 40 (100 EP)	$2000 (100 EP)
Rootzone (horizontal wetland)	0.5–360 kL/d (2·1800EP)	Subsurface flow wetland followed by a vertical filter	UV disinfection is required to reduce pathogen	Disinfection required	4–1600	1000 (2000EP)	

Rootzone (horizontal wetland)	0.5–360 kL/d (2·1800EP)	Subsurface flow wetland followed by a vertical filter	UV disinfection is required to reduce pathogen	Disinfection required	4–1600	1000 (2000 EP)	
WaterPac	2–10 kL/d (20·100 EP)	Biological system – primary settling followed by recirculating media fioltration	Suitable for smaller communities and cluster developments with land available	Restricted irrigation	20–200	10 (10–20 EP) 60 (40 EP) 120 (100 EP)	$500–$2500
KEWT	7.5–1300 kL/d (30·6500 EP)	Primary separation in septic tank, filtered and then evaportranspiration	Further treatment required for water to be used for non-potable water. Stuiable for developments with land avaialble for irrigation	Restricted or subsurface irrigation	200	30–50 (50 EP)	-
Innoflow	10–100 kL/d (50·500 EP)	Onsite primary and biological treatment with centralised effluent treatment (recirculating textile filtrater)	Small diameter sewer minimises exfiltration protecting the environment. Water reuse can be achieved by including further disinfection	Disinfection required	160 + onsite intercept	500 (100 EP) includes reticulation	$400 + periodic maintenance

Notes:

1 EP is defined as 'equivalent person' as 200 L/p/d. Typical operating range describes indicative operating ranges

2 This table is adapted from LANDCOM <www.landcom.nsw.gov.au>

Key	
Suitable for toilet flushing	
Suitable for outdoor uses	
Suitable for cold washing machine tap	

Water treatment technology	Typical scale[1]	Treatment technology	Other information	Water quality suitable for[2]	Footprint (m^2)	Capital expenditure ($000)	Operating costs (per year)
Packaged environmental solutions (ISWTETS)	12.5–100 kL/d (50-400 EP)	Biological treatment followed by membrane filtration			60	350	$25 000 (40 EP)
Water fresh	43–1000 kL/d (300-3000 EP)	Disinfection by a high velocity sonic disintegrator, then Struvite crystaliser coupled with filtration	Higher quality water can be attained with ultrafiltration or reverse osmosis	Restricted or subsurface irrigation	32	-	-
COPA – ReAqua M5R	15–300 kL/d (80-1500 EP)	Membrane bioreactor	Disinfection required		50–180	150–1000	$5800 (80 EP) $21 000 (1500 EP)
Aquatec – Maxoon – Kunota	100–300 kL/d (500-1500 EP)	Membrane bioreactor	Kubota membrane. Disinfection required		20–70	380–452	$15 000–$23 000 (100 kL/d) $35 000–$45 000 (30 kL/d)
Port marine	40–1600 kL/d (200-8000 EP)	Membrane bioreactor	UV disinfection is required for non-potable urban uses		Small	590	$56 125 for 30 kL/d
Ludowlol – Zenon	5–1000 kL/d (50-4000 EP)	Membrane bioreactor	Disinfection required		55–150	50–1400	$30 000 (50 kL/d) $100 000 (200 kL/d)
Veolia	100–500 kL/d (500-2500 EP)	Membrane bioreactor	Disinfection required	Disinfection required	20	540–1400	$0.55/kL $20 000 (100 kL/d) $100 000 (500 kL/d)

Localised development

Localised residential development, eg multi-unit dwellings)

Memcor – Memjet Xpress	100–400 kL/d (500–2000 EP)	Membrane bioreactor	Disinfection required	Disinfection required	50–100	500 (100 kL/d) 800 (400 kL/d)	$55 000 (100 kL/d) $82 000 (400 kL/d)
NuSource Water	50–2000 kL/d (200–8000 EP)	Membrane filtration	This sewer mining process can respond quickly to changing water demands, so little on-site storage is required to cater for seasonal demand, for example, irrigation		30	650 (50 kL/d) 1100 (200 kL/d)	$37 000 (50 kL/d) $84 000 (200 kL/d)
COPA (ReAqua HBNR)	6.2–7000 kL/d (30–35 000 EP)	Biological process – intermittently decanted extended aeration	Disinfection required			200 (30 EP) – 1000 (2000 EP)	$5000 (30 EP) – $25 000 (2000 EP)
Memcor CMF, AXIM	40–>3000 kL/d (200–>15 000 EP)	Membrane filtration	Tertiary treatment processes designed to upgrade secondary effluent. Further front end treatment processes required. Systems greater than 300 kL/d are custom designed		7–13	AXIM 250 (40 kL/d) 350 (500 kL/d) AXIA 45 (500 kL/d) 2500 (3000 kL/d)	AXIM $8000 (40 kL/d) $27 400 (500 kL/d) AXIA $22 000 (500 kL/d) $55 000 (3000 kL/d)
COPA – ReAqua CAS	500–5000 kL/d (2000–25 000 EP)	Fine solids separator (FSS) followed by biological process. Further disinfection is required	The FFS is an excellent solid-liquid separator for processes such as sewer mining		14 for 4500 EP	600	$77 000 (500 kL/d) $150 000 (1000 kL/d)
Baleen	9000–38 000 kL/d (45 000–200 000 EP	Filtration	Further treatment required to attain non-potable urban water uses	Further treatment required	3.5–9	24	$0.65/100 kL

Large scale residential development

A7 Performance of the tested greywater technologies in the WaND research

Table A7.1 *Performance of the five greywater treatment technologues tested in the WaND research project*

	Feed	MBR		MCR		Feed	HFRB		VFRB		GROW	
		Residual	Removal (%)	Residual	Removal (%)		Residual	Removal (%)	Residual	Removal (%)	Residual	Removal (%)
COD (mg/l)	400 ± 50	48 ± 23	88	79 ± 18	80	452 ± 209	111 ± 57	75	27 ± 29	84	139 ± 78	69
BOD (mg/l)	151 ± 23	1 ± 1	99	9 ± 8	94	151 ± 51	51 ± 35	66	5 ± 6	97	2717 ± 44	53
Turbidity (NTU	29.8 ± 8.0	0.2 ± 0.1	99	0.7 ± 1.1	98	47.2 ± 29.2	11.9 ± 13.0	75	2.2 ± 1.5	95	25.3 ± 13.4	46
DOC (mg/l)	53 ± 8	20 ± 2	62	31 ± 5	42	–	–	–	–	–	–	–
TN (mg/l)	5.9 ± 1.5	3.0 ± 2.3	49	3.9 ± 24	34	–	–	–	–	–	–	–
NH_4^+ (mg/l)	0.1 ± 0.2	0.2 ± 0.4	/	2.5 ± 3.5	/	0.3 ± 0.3	0.4 ± 0.4	/	0.1 ± 0.1	67	0.2 ± 0.1	33
NO_3^- (mg/l)	1.1 ± 0.9	1.8 ± 1.5	/	0.7 ± 0	36	1.2 ± 1.2	0.7 ± 0.1	42	0.6 ± 0.1	50	0.7 ± 0.1	42
PO_4^{3-} (mg/l)	0.4 ± 0.4	1.1 ± 1.6	/	0.1 ± 0.1	75	0.3 ± 0.2	0.5 ± 0.2	/	0.1 ± 0.1	33	1.0 ± 0.8	/
SS (mg/l)	51 ± 27	2 ± 2	96	2 ± 1	96	87 ± 65	31 ± 16	64	9 ± 6	90	19 ± 9	78
Total coliforms ms (cfu/100 ml)	$5 \times 10^7 \pm 8 \times 10^7$	4 ± 8	7.1	0 ± 0	7.7	$5 \times 10^7 \pm 8 \times 10^7$	$5 \times 10^4 \pm 10^5$	3.0	$3 \times 10^4 \pm 4 \times 10^4$	3.2	$10^6 \pm 10^6$	1.7
E.Coli (cfu/ 100 ml)	$4 \times 10^4 \pm 10^5$	0 ± 0	4.6	0 ± 0	4.6	$4 \times 10^4 \pm 10^5$	$3 \times 10^2 \pm 8 \times 10^2$	2.1	$1 \times 10^2 \pm 2 \times 10^2$	2.6	$10^3 \pm 2 \times 10^3$	1.6

A8 Contact details of organisations involved in developing the decision support details

Tool	Name of organisation	Contact
Project assessment tool	University of Sheffield	Richard Ashley
Site screening tool	University of Exeter	David Butler
UWOT	University of Exeter	David Butler
Suitability evaluation tool	University of Exeter	David Butler
Demand forecasting tools	University of Leeds	Adrian McDonald
Stormwater management tools	HR Wallingford	Richard Kellagher
Greywater recycling tools	University of Exeter	Fayyaz Ali Memon